Planning and Designing the IP Broadcast Facility

The transition to computer-based technologies and file-based workflows is one of the most significant changes the broadcast and production industry has seen. Media is produced for multiple delivery platforms: Over-the-Air, Over-the-Top, large screen displays, cable, satellite, web, digital signage, tablets, and smartphones. These changes impact all aspects of creation, production, media management, technical operations, business processes, and distribution to end users. Of all the books and papers discussing storage mapping, packet transport, and compression algorithms, none puts all the pieces together and explains where these fit into the whole environment. *Planning and Designing the IP Broadcast Facility* is the first to provide a comprehensive understanding of the technology architecture, physical facility changes, and—most importantly—the new media management workflows and business processes to support the entire lifecycle of the IP broadcast facility from an engineering and workflow perspective.

Key features:

- This beginning-to-end perspective gives you the necessary knowledge to make the decisions to implement a cost-effective file-based production and distribution system.
- The cohesive, big-picture viewpoint helps you identify the differences in a tape-based facility, then how to overcome the unique challenges of upgrading your plant.
- Case studies throughout the book serve as recommendations and examples of use, helping you weigh the pros and cons of various approaches.

Gary Olson is an advisor specializing in the transition of traditional media workflows and business processes. As a designer, he has provided his knowledge to organizations implementing IP and file-based technology. His focus is the adaptation of organizational structure, staffing models, and workflows to implement digital media technology. Gary is a recognized industry leader with practical experience in the analysis, selection, and uses of technology and as an innovator in media technologies and broadcast design. He designed the first commercial television networks for countries in Central Europe, the Caribbean, and South America. Gary holds a US patent in streaming media automation and distribution.

Planning and Designing the IP Broadcast Facility

A New Puzzle to Solve

GARY OLSON

Focal Press
Taylor & Francis Group

NEW YORK AND LONDON

First published 2015
by Focal Press
70 Blanchard Road, Suite 402, Burlington, MA 01803

and by Focal Press
2 Park Square, Milton Park, Abingdon, Oxon OX14 4RN

*Focal Press is an imprint of the Taylor & Francis Group,
an informa business*

Library of Congress Cataloging-in-Publication Data
Olson, Gary (Broadcast engineer)
  Planning and designing the IP broadcast facility : a new puzzle
to solve / Gary Olson.
    pages cm
 1. Webcasting.   2. Internet radio broadcasting.   3. Digital
media.   4. Communication and technology.   I. Title.
  TK5105.887.O47 2014
  006.7'876—dc23
  2014019007

ISBN: 978-1-138-79895-3 (hbk)
ISBN: 978-1-138-79896-0 (pbk)
ISBN: 978-1-315-75630-1 (ebk)

Typeset in Minion
By Apex CoVantage, LLC

Printed and bound in Great Britain by
CPI Group (UK) Ltd, Croydon, CR0 4YY

Dedication

This is dedicated to my wife Ellen who has given me unwavering love and support through all my changes and adventures.

Contents

About the Author

GARY OLSON is a Technology Architect and Advisor for Digital Media Strategies, who has developed a wide range of innovative broadcast, digital media, content management, and information products and services.

Gary has designed broadcast centers for major US networks and for emerging countries introducing their first commercial television channels. His designs have always incorporated a "future proof" philosophy, using innovative technology solutions.

He is an evangelist for the transition to an IP in the broadcast and production industry, creating media management system designs.

Gary advises on the best use of technology in the migration and transition from traditional media technology to digital platforms for broadcast, broadband, wireless, portable multimedia, and On-Demand media.

His substantial experience in working with project sponsors and stakeholders helps them understand the changes associated with digital media platforms and file-based workflows.

Gary has advised major US and international broadcasters, corporations, and institutions on new digital production and distribution. He is a recognized industry leader with practical experience in the analysis, selection, and uses of technology and as an innovator in media technologies and broadcast design.

Acknowledgments

I need to thank Peter Cresse and Andrea Cummis for pushing me to write my first whitepaper that started this process. I hadn't thought of myself as a writer. The idea for the book came from conversations and discussions with so many people in the industry about all the end-to-end solutions that didn't make a complete system. Capturing that into the puzzle concept was in collaboration with my niece Jordan and turning the concept to illustrations was thanks to Paul and Judy. There are so many friends and colleagues I need to thank for giving me their knowledge and support on this adventure. I want to acknowledge TBC Consoles for allowing me to use their images to demonstrate some of the concepts. Other images are from my own projects. All of the standards and protocols shown are in the public domain published by all the different Standards organizations.

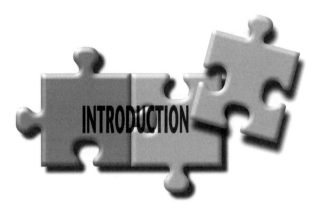

INTRODUCTION

The transition occurring in the broadcast and media production industry is a profound **game changer**. It is one of the most dramatic and significant changes that impacts every aspect of creation, production, management, distribution, and monetization. It is more than technology, it is changing the way people work, the interaction between business units and how content is consumed.

The IP- and file-based architecture and the new concepts in workflows are very different from the proprietary tape–based SDI technology and legacy workflows that have been inherited from the early days of television.

This book examines the spectrum of differences in the planning and design of an IP- and file-based infrastructure from a technical, operational, and architectural perspective. Whether planning and designing for a new facility or updating and evolving an existing one, there will be a significant impact on the lifecycle of media. The changes in the complete lifecycle also have had a considerable effect on the workflow, relationships between business and broadcast, plus the change to business processes.

The book focuses on the "beginning-to-end" entire architecture and media lifecycle, including all of the workflows and processes—rather than on the over-used statement of "end-to-end." There is much discussion in the broadcast and production industry that the complete architecture is an end-to-end problem to solve. However, there is no silver bullet or single solution that solves all the issues of integration, business, and workflow or technology associated with the entire broadcast and production lifecycle of media and its value chain. The design considerations need to start from the very beginning (acquisition) and continue right through to the end user experience (distribution and delivery).

The goal of this book is to provide some useful information and guidelines to the broadcast engineer, system designer, and technical management of the changes to the technology architecture and workflows within the IP- and file-based architecture that impact the entire lifecycle and value chain of media.

There are a lot of facets to this, so to assist with a little structure, the book is organized in a way that will provide a comprehensive understanding of all the aspects that need consideration when designing an IP ecosystem and transitioning into IP. The book starts with an overview and then each chapter delves deeper into each of the different areas.

There have been significant milestones in the growth and evolution of broadcast and production technologies and processes. This examines the profound and substantial changes that the transition to IP is creating. It can certainly be looked at as a new puzzle to solve.

one

To begin solving this new puzzle, first it is necessary to identify all the pieces. Then they can be assembled into the complete new picture. There are many familiar elements to this new architecture and the workflows.

- Ingest, Acquisition, and Capture
- Workflow and Business Processes
- Media Management
- Technology Infrastructure and Engineering
- Transmission, Delivery, and Distribution
- Facility Planning and Design

This new picture will provide an understanding of what goes into the file-based architecture in the IP ecosystem.

The core infrastructure has changed substantially in broadcast and production technology in the world of IP. The design and implementation of IP- and file-based technologies and infrastructure also has to cover new workflows, processes, and facilities design.

One of the critical puzzle pieces is the integration of business units. The introduction of new workflow integration between business units and media is the need to work more tightly together. This means that many of the operational silos that existed before are now being broken down. Traditionally, media and business units operated largely autonomously from each other, as in silos. This "silo" organizational structure existed even within each of the business and media groups.

Within the broadcast, production, and technical operations, finance, business, and legal departments, they all operated independently. Now in the new business model, everyone in the organization needs, will, and can access the media and metadata. Metadata in the context of media management has critical implications and is core to what the IP- and the file-based world is all about.

One of the many changes that occurs is that the relationship between technical operations, engineering, and business demands tighter integration. The broadcast center has now become a media data center. Governance, once the province of enterprise IT, now plays a significant role in the rules and policies that control and manage workflows and processes of media. These rules and policies are new and different when applied to what IP means and how IP now exists in the broadcast space and beyond. Program delivery is across many platforms and devices, in a time when the consumer watches media on multiple devices and platforms (phones, tablets, and PCs).

This significant change in the way media is delivered impacts production as each platform can have a different format and its own delivery specifications.

Not only has the technology changed, but there has also been significant impact on the design and planning of the physical plant—the actual bricks, mortar, and electrical and mechanical infrastructures.

These new puzzle pieces fit together into a completely different picture, in terms of the changes in technology, workflows, the design of the facility as well as the design of the infrastructure.

That new picture is what this book addresses and represents the entirety of the now-changed landscape of producing, managing, and delivering media.

New Terminology

IP is a term that is used to describe many different things. In the technology world, by definition, it started out as TCP/IP, which means: Transmission Control Protocol (TCP) and Internet Protocol (IP). Then it was shortened to just IP. TCP is now only one of a few types of transport protocols.

In the media industry, the transition from tape-based or traditional baseband or SDI technologies to computer-based or "IT" technologies has been coined as IP. The term IP is now used as a more generic definition used to describe the encoding of media using an application running on a server that is transported over a network into a storage environment. It is used to describe how information moves within a media organization's business and media processes across the network architecture. The media industry is using the term IP as a modest differentiator to the IT industry, which uses IP to describe a protocol. The media industry has adopted "IP" to describe more than a protocol or a technology. It refers to the overall environment, including computers, applications, middleware, servers, storage, and network—plus all the workflows and business processes.

In the media management ecosystem, IP is the way media moves and how different systems and subsystems can be connected. This infrastructure has to be designed to support the different layers within the IP architecture, and there are new tools known as middleware to integrate applications. These tools address many of the layers, such as command and control, interfaces between systems, media transport, production, metadata in its many formats (e.g. descriptive, administrative, and structural), management, communications, and distribution.

What does the entire architecture look like from beginning to end? And what are the changes that have taken place? Media management includes the business processes and workflows. In many ways the lifecycle of media itself has not changed, the ecosystem and technology architecture have. The value chain has evolved to more closely integrate many of the processes that were either manually intensive or in systems that did not communicate with each other.

The book will explore the changes to the core processes in the new lifecycle: Creation, Management, Media Movement, Handling, Retention, and Delivery.

Figure 1-1 shows at a very high level the core processes that represent the entire lifecycle of media movement in the IP world.

When raw media is first recorded (ingested, acquired), it is considered *essence*.

As we **ingest** the **essence** and add **metadata** through the **logging process**, the media becomes **content**.

Media is acquired on flash drives, Solid State Drives, hard drives, and optical disks, or it can be transported first as a stream then captured to a file. As SD/HD-SDI is brought into the new *ecosystem*, it gets encoded to IP as a file and then transferred to storage. Once in the storage environment, it enables all users to have access to, edit, distribute, archive, and manage the media.

Studio and field production has changed. For field or remote productions the media can now be viewed immediately, rough cut editing can be performed in the

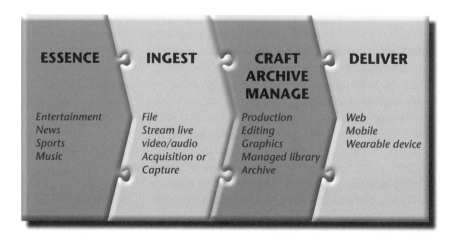

FIGURE 1-1

field, proxies can be posted online for review and approval. Many production and business decisions can be made instantly.

Computer- and server-based systems have changed studio production. There is automation for camera control, video switchers are more powerful with built-in image storage, multi-channel 2D and 3D effects, and clip players for animations. The studio is a virtual set, tied to the camera robotics and lens telemetry. All of this is based on the IP infrastructure. The production consoles are now work surfaces connected via the Ethernet switch topology. Media moves between systems over the same network as command and control.

Craft production, editing, and graphics are all computer-based and add layers of complexity with metadata, media management, and version control.

A different delivery format is required to send content to a phone, tablet, or computer as a stream or a file.

The term IP covers a fairly broad spectrum of meanings, not all of which are related to broadcast.

New Vocabulary

The broadcast and media industry like most others has always had its own lexicon complete with acronyms and abbreviations. The transition to IP- and file-based technologies, workflows, and processes has brought with it new and exciting acronyms and abbreviations with shiny new terminology and a whole new vocabulary. Even the word "vocabulary" has a new application.

In the production world, the term "*shoot*" is still used while "*film*" has been largely replaced by "video," the same applies to how we "*record*"; now it's "*acquire or capture*" content "*and ingest*" to a "*file or stream*" project, script and location notes are now "*metadata that is logged and tagged.*" Here again, and as mentioned earlier, the raw material is called "*essence,*" and once it's associated with "*metadata*" it becomes "*content.*"

SD/HD-SDI is "*Encoded*" and "*files and streams*" are "*Transcoded,*" "*Transmuxed,*" for ingest. There are multiple formats, and for production and craft, the audio and video needs to be "*Embedded,*" "*DeBedded,*" "*Muxed,*" or "*DeMuxed.*" Then for delivery and distribution, they are "*decoded*" or once again "*transcoded or transmuxed*" into many different formats before they are "*spliced*" and "*groomed*" before being "*uploaded or streamed*" to a "content distribution network (CDN)" or "cloud" using a "Content Management System (CMS)."

Editing is done on a "*reference timeline*" using "*proxies or a mezzanine resolution*" and is only "*rendered*" to high resolution for finishing and delivery.

When SDI was introduced there were new tools for Measurement, Test, and Monitoring that provided better information on the quality and integrity of audio and video. The Waveform Monitor and VectorScope for video expanded to include Jitter, Gamut, and Eye Patterns. For audio, Level and Phase metering has expanded to Sample Frequency Accuracy, Channel Compliance, Phase-Lock, Fs Jitter, and Bitrate. File- and stream-based media has new quality control test and measurement tools and analytics. The new measurements are *"bitrate error," "CRC checking," "packet loss detection," "frame rate,"* and *"GOP length"* to list a few. And now *"Quality of Service"* and *"Quality of Experience"* are the new metrics.

Information about media is now *"Metadata"* or *"Data about Data."*

Metadata, too, has its own terminology that includes a *"controlled vocabulary"* with terms such as *"taxonomy"* and *"ontology,"* with *"faceted"* and *"contextual"* searching.

Workflow, Processes, and Integration

The lifecycle of media begins with the creative process and (essence) acquisition. Media workflows, business processes, and systems have become more tightly integrated.

Once the *essence* together with its associated *metadata* has been acquired, ingested, and logged (now "Content"), it must be made available and accessible over the network and throughout the organization. Various business units, such as craft production, media management, library, and business operations—including legal, finance, marketing, rights management, and others—must all work in parallel and in concert to turn this content into an asset, something that has value, is managed and tagged, and can be tracked.

The entire organization has to create and integrate metadata, and also set the rules and rights of the asset. This will position the asset for its journey as it is prepared for distribution and monetization.

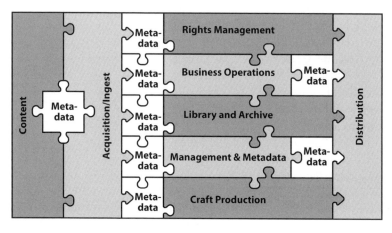

FIGURE 1-2

While the craft of broadcast and production is the same, the technologies and processes have changed a lot. They now include:

- Acquisition/Ingest/Capture
- Craft Production
- Editing
- Graphics
- Virtual Sets
- Motion Capture
- Media Management
- File Movement
- Library & Archive
- Metadata
- Technology Integration – IT
- Test and Measurement
- Stream Monitoring
- File Error Checking
- Network Monitoring
- Technical Operations Workflows & Processes
- Automation
- File Movement
- Business Integration
- Legal
- Finance
- Marketing and Business Intelligence
- Digital Asset Management (DAM)
- Digital Rights Management (DRM)
- Controlled Access, i.e. subscription
- Protection, i.e. Watermarking
- Multiple Delivery and Distribution Platforms
- Over-the-Air (OTA)
- Over-the-Top (OTT)
- Cable, Satellite, IPTV
- Broadband
- Mobile and Wireless
- Gaming Console

Business Integration

The business or enterprise side of a media organization has operated on an IT backbone for many years, using a variety of applications (e.g. office productivity tools, customer management, accounting, databases, etc.), servers, storage and network systems. Now the broadcast and media production side of the organization has moved to a similar backbone, but with the additional considerations and requirements such as Quality of Service (QoS), low latency, security, and access. In this new relationship, the technical operations group supports media operations and ties the business units to media services. There is a need for close collaboration between enterprise IT and broadcast IT and both must respect the differences in technology considerations that result from this change.

As the media is delivered to so many different platforms, there are new concerns about access and protection.

- Over-the-Top (OTT)
- Over-the-Air (OTA)
- Video On-Demand (VOD)
- Streaming—Phones, PCs, tablets, or game consoles

These are managed through metadata and media management tools.

Technology integration between media and enterprise IT needs to protect both the business network and the media network.

The enterprise and broadcast networks now intersect and media management crosses over to the enterprise while both networks need to be protected from each other and from the outside.

In the media IP- and file-based ecosystem, managing the library and archive is different from just storing tapes on a shelf while using yellow pads with segment and timecode information; and tracking them in a database. A digital library is a repository of media that is dependent on metadata, asset management tools, and a storage architecture that includes removable media for long-term archiving. The metadata is associated with the media by using asset management tools and databases.

Technology Infrastructure

The core infrastructure is now comprised of applications, servers, storage, and network. When planning and designing any facility it is critical to anticipate growth. The core technologies in the IP infrastructure are different from tape-based SDI technologies and have other considerations:

- Extensibility
- Scalability
- Sustainability

Extensibility: This is a design principle where the implementation takes into consideration planning for future growth.

With technology changing as rapidly as it does, it is essential in the planning and design of a facility to assure that extensibility provides for and *anticipates* future growth.

Scalability: The ability of the hardware and software systems to expand to handle the growth without requiring replacement. This growth can be an increase in the volume of material acquired, produced, delivered, or archived. It could be the number of users needing access or an increase in delivery platforms. What about the growing amount of proxies and metadata?

Scaling, growing, or expanding in the IP ecosystem is accomplished by adding servers, network, storage, and applications. Are there enough network ports? Is there enough bandwidth? If more storage capacity is added, what about throughput?

Sustainability: Maintaining hardware and software is all about service contracts, software updates, and patches. Sustainability includes keeping spares or redundant systems as hot spares or backup.

Total Cost of Ownership

Planning a new facility or making a significant capital investment in an existing facility, one of the critical elements of budgeting and decision-making is the *total cost of ownership* of equipment and infrastructure. It is important to know not just what it costs to build, but also the cost to operate and maintain. *What does it take to run*? For the IP architecture, there are a number of new considerations.

Facility Infrastructure—Physical Plant

Infrastructure is about more than just the technology. It is also about the physical facility, and with that there are a considerable number of changes to consider in planning and design.

It is generally accepted that building a new facility is always easier than upgrading an existing one. Many of the considerations are the same, but it's always a little easier to build from scratch, rather than modify existing infrastructure. This is as true with the physical aspect of the project (bricks and mortar) as it is with the technology. It's also much easier for business continuity if the upgrade does not involve working around live systems. It is something like changing the tires while the car is moving.

When thinking about making building changes to accommodate IP- and file-based systems, space planning is completely different. New control surfaces do not require the same proximity to a mainframe for the proprietary connection requirements because they are now IP and connected over the Ethernet network. Even the server controls—screens, keyboards, and mice—are accessed using extenders with a switch matrix that allows a single screen, keyboard, and mouse to operate multiple servers from many locations.

Core space planning includes:

- Room Adjacencies
- Spatial Allocations

- Room Interdependencies
- Ergonomics

Production control rooms no longer accommodate as many dedicated control surfaces or have as many individual monitors by using multi-viewers, so they're more efficient.

A typical console and room layout still has the same design considerations. These include:

- The quantity and type of control surfaces
- How much support equipment
- How many screens or windows on the multi-viewer
- Ergonomics—the placement and proximity of primary control surfaces

Heat and Power Loads and HVAC Design

Power-wise, spaces today need a lot less power, and ergonomically they are easier to manage and operate. In the control rooms there are mostly control surfaces not the equipment frames or servers. In the equipment room the production systems have more of the features and functionality embedded in fewer physical devices. This significantly reduces the heat load. An example of this is the production switcher, it is a composite of functionality, encompassing a video switcher, still store, digital effects, and clip animation player.

Operating Costs—Space/Power/HVAC

The amount of physical space required factors into the total cost of ownership, including basic overhead costs such as space, power, and mechanical systems. It is a reasonable assumption that if there is less space in an IP-based architecture, it's possible that the cost of overhead is lower. The amount of mechanical systems that are needed for the control rooms and support spaces is reduced significantly and that becomes a reduction in capital costs that can also translate into potential cost-savings on power.

In the IP facility, there are considerable changes to all control room designs. There are fewer control surfaces because of the greater feature sets and consolidation of functionality built into the hardware and software. This creates a significant change in console layouts.

Media management and metadata (for description, business, and automation) can be entered at the same time that craft production is taking place.

The media manager and automation system are applications that handle the transport of the file as it moves through the system. As content goes through the encoder and the media manager, and then into production and

FIGURE 1-3A

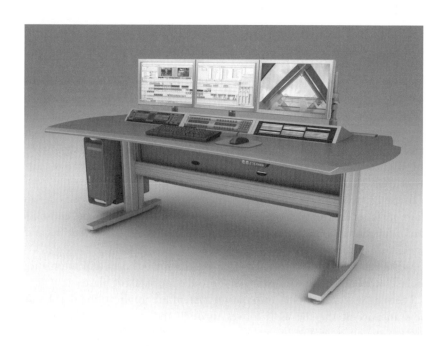

FIGURE 1-3B

distribution systems while insuring a copy goes into archive, automation plays a much larger role.

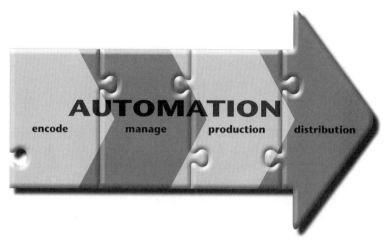

FIGURE 1-4

Changes in Workflow

The interaction between business units and the handling of media has changed many of the workflows and business processes. Business models have also significantly changed for creating and distributing media. As media moves throughout the entire infrastructure, it needs to be accessible to all departments/units including production, distribution, library, legal, finance, marketing, and business intelligence. These groups need access to the media to input and export metadata and to ensure that the appropriate business rules, policies, and permissions are in place. This metadata will protect the content as it is distributed into the marketplace.

Beginning-to-End Planning

'Beginning-to-End' is the concept of planning and design that must include business operations, technical operations, and workflow management.

Workflow is a term often used and sometimes abused. It is defined as a series of connected activities and processes. For example, editing workflow is the process of editing plus all the different pieces and parts associated with editing. Accounting workflow relates to all the things associated with payables, receivables, and accounting work.

There are many workflows within the world of media, technical operations, and their association with business operations. Some new processes are made possible by adding technology and achieved by changing the way people work.

During **acquisition, essence** is **captured** and **ingested** using encoders and transcoders. In the **craft** workflow there is **Remote** and **Studio** production, **editing**, and **graphics**. These are all workflows that change in the IP ecosystem. Each of these processes needs simultaneous access to the same media files in the shared storage over the IP architecture.

Media Management

Media management means many things to different people, and it is often confused with only being asset management. Media management is the entire collection of processes and technologies that handle media throughout its lifecycle.

Historically, media management was considered a traffic and library function. Now that the business process is tightly integrated with the handling of media, <u>**it could be easily said that the entire new lifecycle from beginning to end is called media management**</u>.

Asset management, specifically digital asset management, assumes that we have something digital, we've applied some value to it, have started calling it an "asset," and that this asset needs to be managed. The term for raw media is *essence*. A video that is captured, given a title, and possesses a reason for being (value) becomes "an asset." By assigning a title and other descriptive data to the asset now allows it to be indexed, organized, and managed in a way that is easier for users to search and find it without needing to know very much about it.

The **asset management** system is the master repository for the **metadata** that has to be **logged**, indexed, and managed. Metadata has to be associated with the media for the media to have any value. Media without metadata cannot be managed or monetized. Asset management assumes that enough metadata has been applied to the content that has become an asset.

Asset management is the process of attaching useful information to media that allows the asset to be tracked and searched as it moves throughout the value chain. This includes saving a copy to the library for **archiving**, thereby enabling access for future use, say in a week, a year, or a decade. It's important that when an asset is requested again, there is enough information to tell what it is, why and when it was made, and who has the rights to use and access it.

The media management system creates and maintains the rules, vocabulary, and structure for the metadata that are applied to the asset in order for it to be useful. Files are archived in a digital library so users will be able to access them in the future. For example, a sports program with a great catch, an amazing pitch or a significant news event needs to be tagged with metadata (i.e. keywords, who, what, and where) so the content can become searchable and retrievable.

Metadata is not only used for management and automation. It has an important use in driving electronic program guides (EPGs), enabling end users to find their programs. It's also increasingly used to drive monetization as part of new business models for second and third screen revenue.

Assuming all media is being archived, what format is it archived in? If it's going to be archived on removable media, there are decisions to make. Is it optical disk, flash, digital tape or does it stay on spinning disk?

An Example of IP Architecture for Media Management

Imagine that there are different LANs, with different application servers, and different users that will access the information across various networks. On the broadcast network (BLAN), user workstations manage tasks on application servers with systems that move media around, handle, ingest, archive, and play media. There are database servers to manage the metadata and video servers to handle the audio and video. These applications can be on dedicated servers and/or on virtual servers (i.e. VMs or virtual machines).

The next questions are about the storage architecture:

- Where are the video and audio coming in from?
- Where is it landing?

Is there a remote access storage system to receive incoming media over FTP and is there proxy storage that handles the media formatted for searching and browsing? There should be separate storage for craft and production so work in progress doesn't consume valuable real estate on the main media storage. The interaction between the internal networks as well as allowing access to outside users is governed by a set of rules and policies.

Another design component of media management is **automation**, which is how the systems and subsystems interact with each other. Media has to move seamlessly throughout the IP infrastructure among different application servers, over the network through various storage systems, from ingest to play-out. In **distribution**, the automation is responsible for controlling the delivery to a number of distribution channels each with a different requirement. Content is sent to a transmitter for Over-the-Air and to a cable or satellite provider for traditional television distribution. And it is delivered to a content distribution network (CDN) for delivery to the web, a tablet, a phone, and Over-the-Top. Automation is a bigger concept than just play-out, and there are new tools to manage automation.

Therefore, to manage media and how it's handled, the whole architecture must be viewed holistically.

Engineering in an IP world

The role of the broadcast engineer has changed dramatically. There are now new skills and knowledge required to maintain the IP media environment. The long-standing broadcast engineers' philosophy that media needs to be the highest quality with the highest service priority has NOT changed.

Quality Assurance and Quality Control

A broadcast center needs to be designed with the ability to ingest all these different formats and insure their integrity, while making them available for inclusion in live programs or to be used as captured content in craft production.

One of the greatest challenges with all these different acquisition and contribution formats is *quality control*.

Quality control begins in the field before a file is transferred. As live streams are transported, they too need evaluation and analysis. The Waveform monitor, VectorScope, audio metering, and spectrum analyzers are still part of the engineer's tool set with SD/HD-SDI. In the IP environment, there are new ways to monitor and measure the quality of the signal in real-time. If an artifact needs to be corrected or fixed, it shouldn't have to wait until a recording is complete (when it's too late). In the IP and file world, there are new tools such as bitrate analyzers, packet loss detection, and bit error detectors for file and stream analysis. There are also different types of analyzers to determine the integrity of the network, such as bandwidth, port speeds, VLAN traffic, and many more. The configuration of the network has a significant impact on both performance and quality control for media movement.

Another thing to monitor is the integrity of the metadata in the database. When metadata moves along, it moves between different databases and across systems.

In the IP architecture, quality control also applies to the applications and servers. As files and streams move back and forth across IP networks, bandwidth becomes a major consideration.

Which of these must be prioritized? What about Quality of Service (QoS)? Network monitoring is required to ensure it's up all the time, in terms of the integrity of the data and storage monitoring. Disks also require routine maintenance.

The acquisition process captures content to removable media encoded from AES/SDI, and then brings it in as a stream and/or file. It lands on a server and is then moved to storage, where it's managed. During the evolution to SDI, the audio was embedded into the video for transport and distribution. In the transition to IP, the next step of the encoding process is to embed audio, video, control, and management data into the file or stream using different acquisition, transport, media storage, and distribution protocols. *This all begins with acquisition.*

During acquisition, the focus is on what has changed with how media is captured and ingested, and that many of the rules are different from before. As a result, there are numerous format and codec decisions to be made.

IP has different tools and metrics for use in quality control, format control, and metadata management. These tools assure that all of the critical elements are present, so that when production, business, and distribution people look for media and metadata they can access it easily based on the policies and rules.

Broadcast now shares an IT-centric world that uses Ethernet network switches, servers running media applications, and complex storage systems that all are handling high bandwidth streams plus large bulky files. The concept of quality control monitoring with test and measurement hasn't changed; there are just new parameters that need monitoring. Some of these new parameters deal with priorities in the network—like bandwidth utilization and quality of service (QoS)—and some are more subjective, like quality of experience (QoE). These parameters are measurable against a set of metrics determined by the engineer.

In the IP environment, there are new challenges. One of these is latency, previously known as delay. Delay has always been a serious problem in media processing and delivery. In the IP-based architecture command and control is on the same network as media transport. If the network is not carefully configured there can be latency. One example of latency is when the play-out system requests a program to air, and the network traffic introduces a delay that prevents the request from arriving; there will be dead air until the request is executed. There are a number of areas within the flow of digital media where latency can be introduced. One of the more noticeable aspects of latency is the synchronization of audio and video within a program element, also known as lip sync. In the world of digital media, there is an interesting phenomenon where broadcasters have compromised their level of quality and allowed a certain amount of acceptable latency when it comes to lip sync.

In the broadcast world, there is no such thing as "off the air." All systems have to be up 24/7/365 and they have to work. Routine and preventative maintenance is something that has to happen in real-time while systems are running. This would be something like changing the tires while the car is driving on the highway.

IP networks have to be protected and security is both of deep concern and also presents many challenges. There is internal security, external security, and operational security. The business side wants to protect their enterprise data because large media files could slow the business down and congest the network. Broadcast engineers have to protect their network because they have to be on the air. Therefore, bandwidth management is critical in the IP infrastructure because it ensures that nothing prevents the media from moving. Bringing media into the system from removable media can introduce viruses or Trojans.

There are differences in providing technical support in broadcast and production vs. enterprise. Solving technical problems is *not* calling the help desk! The broadcast engineer *is* the help desk and has to be equally responsive to IP issues and staying on the air. It's no different than SDI.

This overview has taken a high-altitude look at the entire media lifecycle from *beginning through to the end.* It introduces what constitutes an IP infrastructure and the multitude of elements that need to be considered when designing, planning, and building an IP infrastructure.

There are new considerations in the operation of an IP-based ecosystem that range from the acquisition process to media management, craft production, business unit integration, library, archive, and retention. There are accessibility issues and a new breed of multiple stakeholders and business units that search through and access content. Media is delivered to all platforms in different forms.

two

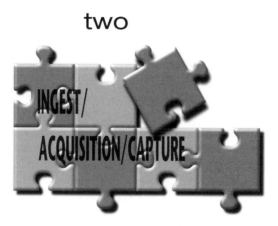

There are many business processes that are part of the lifecycle of media and media management. The creation or acquisition process begins the media handling part of the media management process. Acquisition is the capture of essence and adding metadata to make it into content so it can be ingested into the IP- and file-based workflow. The acquisition piece of the puzzle is going to address the technologies, workflows, and processes related to remote and studio production, as well as all the various forms of contribution that bring media into the broadcast production facility.

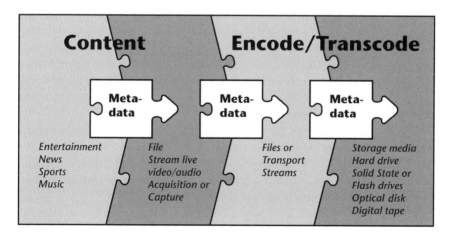

FIGURE 2-1

The lifecycle of media actually begins before any media is actually created. Early in the process contracts are signed, distribution rights negotiated, and production is planned. Each of these processes creates metadata. Next in the media management ecosystem is the acquisition process, which includes studio and remote production plus contribution. This is the critical stage where format and removable media type decisions are made prior to ingest. Essence is the term used to describe the raw audio, video, and still images. It is acquired through cameras and microphones and then encoded into files or transport streams. This is also the point where tagging and logging of metadata begins.

When the metadata is associated with the essence it becomes content that can be created as a file or stream. If it is created directly to a file and transferred to storage, is the storage removable media? There are a number of different kinds of storage media: a hard drive, solid state or flash drives, optical disk or digital tape. If it is encoded to a stream for transport then there are different formats for streams before it is captured to a file. There are different formats used for sending content from the field as remote production or as contribution depending on if it is a file or a stream. When the content is captured to file, the files need to be transported back to the broadcast center. If there is enough bandwidth available, the files can be transferred (uploaded) to the broadcast center or, if not, then they need to be transferred to removable media and then physically transported (courier or hand carried).

If the content is going to be transported as a stream there are also a number of options and decisions. They can be sent in their native SDI form or encoded into multiple kinds of IP streams, such as ASI, MPEGTS, JPEG2000, MPEG4 H.264, HTML5, and HTTPLive. Most people think that by sending SDI Uncompressed (SD-SDI is 275Mb/s and HD-SDI is 1.4Gb/s) is using an uncompressed transport circuit and that there is no compression. Actually most video circuits are still based on some form of compression so while the input and output of the circuit may be SD/HD-SDI, it is still compressed. Once it is received at the broadcast facility, it is encoded to a file and transferred to storage. If the SD/HD-SDI is encoded to a stream in the field, then the type of transport changes, fast Ethernet circuits are 1Gb/s up to 100Gb/s. The stream will still be received by servers, captured as a file, and transferred into the media storage system. It can also be passed directly through to distribution as a live stream. When it is live studio production, it is encoded to a file into media storage and is also streamed live to distribution. Streaming media live for distribution is a different workflow and process from delivering files for distribution.

Whether the content is a stream or a file, quality assurance is critical, but the methods for handling the two differ significantly. This will be discussed in greater detail in the engineering chapter. As we are capturing the essence, there is a logging and tagging process to enter metadata that is a core component in media management, whether for a stream or a file in the IP ecosystem.

The figure below shows when media (or essence) is imported into an IP infrastructure, the SDI video and/or AES audio must first be encoded. The encoding process involves multiple format options for files and streams.

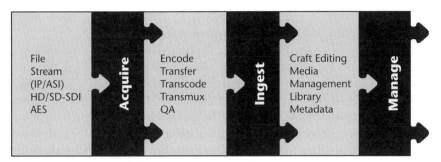

FIGURE 2-2

Acquire, Ingest, and Manage

What used to be called baseband in analog is now called Wideband (SD) or Ultra Wideband (HD) in SDI and now there is UltraHD (4K) which is four (4) times the number of pixels than HD-SDI, each of these are the terms used to define the content as uncompressed. In the acquisition workflow, the SDI and 4K has to be captured and encoded to a file or stream.

If the content is already in file form, then the ingest process is handled by file transfer and if the file format is not in the house standard format then file conversion is needed.

This now becomes a *transcoding* or *trans-muxing* process. What's the difference between *transcoding* and *transmuxing*?

Transcoding is the process of changing the format of audio or video through a partial decode and re-encode. This is considered a very lossy process.

Transmuxing is the process of changing the format of a video or audio file while preserving some or all of the streams from the original file. Transmuxing is a more efficient way to convert a file, and involves little or no loss.

If the content originates as SDI, then it needs to be encoded. The encoding process and format decision includes a decision on compression. If the content is encoded to a stream then there are multiple decisions. What is the transport format; and then as it arrives in the server as a stream, how will the stream be captured to a file?

Compression

When media is compressed, sophisticated algorithms are used to calculate what bits can be removed during the encoding process without affecting or minimally affecting the quality of the content when it is decoded. There are many different encoding schemas. MPEG2 follows an I Frame and Long GOP (Long Group of Pictures) format, JPEG2000 is based on the Discreet Cosine Transform (DCT) methodology. MPEG4 H.264 follows closely to MPEG2 and the open source formats like ON2 follow a number of different compression formats.

The basic rules are simple, the higher the compression rate, the more content gets removed and the more apparent it is when viewing and listening to the video and audio. Lossless compression suggests that while there is content removed, there is no perceptible change to the quality of the content. Lossy compression suggests that in the compression process there will be discernible quality loss in the content. The compression process evaluates key frames, groups of pictures (GOP), white and black spaces, colorimetry, motion, and silence and then the algorithms make determinations of what can be reliably reproduced without compomrising the quality of the content during the decoding process. In the decoding process, one of these techniques is called interpolation. This is mostly used in sports slow motion for instant replay. What is interpolation? If there is a gap between two frames, the algorithm makes an assumption based on the frames that bracket the gap, the before and after frame. It creates a filler that replaces the gap. This has caused some interesting discussions when officials using slow motion instant replay are making judgment calls for goals or infractions. For example, if the replay is to determine if a player is out of bounds and the officials look at a replay, if the point where the player's toe is too close to the boundary is in one frame and the next frame is missing, but the following frame is on the line, the algorithm in interpolating the missing frame could put the toe in or out which can change the outcome of the game. If the interpolation is not done, then the game official has to make the call based on essentially missing video. It presents a challenging question of allowing the algorithm to add a best guess.

When the content is decompressed, the algorithms track the compression schema and, during the decoding process, replace missing content by making theoretically intelligent decisions where either duplicate pieces of the retained content or interpolative assumptions can be used to restore the full content without any perceptible change or loss.

Since each compression process is unique and there is no catalogue of what the first compression process removed, the next stage of compression removes something completely different. This means that each time media is compressed, encoded, decompressed, and recompressed, it is impossible to get back what was originally compressed—meaning what was removed. Therefore, each compression process is forced to remove new things. In the world of digital media theoretically there is NO loss in transfer or copy—or maybe not. Each iteration of compression adds more loss. This is called *concatenative* loss, a fancy word, but it still just means loss. This is essentially the same as degenerative loss during the tape duplication and transfer process.

Quality Control

The concern about quality control has not changed. There are new tools and quality control processes to maintain consistency for a stream or a file. One of the issues that file-based production introduces is that before a quality control tool can analyze a file, the file first needs to be fully rendered. File analyzers look at the integrity of the file and a stream analyzer looks at the packets and compares them to the parameters of the protocol selected. Speed becomes an issue here. If a program is 1 hour long and the file analysis is in real-time and takes an hour, the program is not available for craft or distribution for 2 hours. The manufacturers of the testing tools are working to improve this by having the process happen faster than real-time or check the media as it is being written to a file.

Manage

Once media is encoded as part of the ingest process, it moves to shared storage based on the architecture. This may be production storage or centralized storage. Once in the storage infrastructure, it can be concurrently accessed through editing, management, library, distribution, and metadata processes.

In production and contribution, which storage and removable media format to use is just one of many questions. How many formats are there?

- Optical disk
- Flash or Solid State Drives
- Digital tape
- Hard disk

Cloud storage is not a unique type of storage; it is a place. Instead of building a storage architecture in the broadcast center it is built in a separate datacenter. It depends on the same storage technologies, it is a different way to access, store, and manage media.

Streams have their own formats. Now is a good time to remember that SDI is the acronym for Serial Digital Interface. There is Standard Definition (SD), High Definition (HD), and Ultra High Definition (UltraHD; 4K). Each of these is a protocol based on published standards or is it a standard based on published protocols? This is a topic for another chapter! SD/HD-SDI are MPEG2 formats that are typically considered uncompressed in the broadcast architecture.

- HD-SDI can be multiple formats i.e. 1080i, 1080p, 720p, or 480p. These numbers determine vertical and horizontal resolution, aspect ratio, pixel aspect ratio, scanning and frame rate of the content.
- The introduction of 4K as UltraHD raises the horizontal and vertical, aspect ratio, pixel aspect ratio, and frame rate which all translate into a considerable increase in bitrate. This introduces challenges in production, storage, transport and delivery for distribution including to displays.

In the IP ecosystem, SDI can be encoded to a broad spectrum of formats. One of the primary considerations with respect to selecting a format is its intended use or delivery platform. This can be production, archive, OTT, mobile, or web.

In addition to encoding, media can be ingested directly from other IP sources as files or streams and may require transcoding. Managing this process starts getting complex. It may be packaged into a container to move across systems and devices throughout the architecture. There are a number of standards and protocols when media moves through the IP environment whether as a file or wrapped in a container:

- MXF – is the SMPTE standard for craft production and archive
- GXF – is primarily used for delivery
- BXF – has superseded GXF with more efficient packaging
- ProRes – is a proprietary Apple technology

These are all container formats. The file itself within the container is a different format. Some of the file formats are MPEG2, MPEG4, JPEG2000 and Quicktime which are usually associated with the video, and AAC, WAV, and PCM are the formats associated with audio. The containers are a way the audio and video can be packaged together in an interoperable transport file to move between systems. If the media is a stream, it can be delivered over different transport systems using SDI, IP, or ASI. Media can be transported compressed or uncompressed and both files and streams are handled in the same way. SDI is considered uncompressed. There are different formats, standards, and protocols used to move media.

All of these are transport protocols:

- SMPTE – SD-SDI SMPTE 259M
    - HD-SDI SMPTE 292M
- ETSI – ASI- TR 101 891 Technical Report
- MPEG2 – ISO/IEC 13818
- MPEGTS – ISO/IEC 13818-1
- MPEG4 – ISO/IEC 14496
- MPEG4 H.264 – ISO/IEC 14496-10
- JPEG2000 – ISO/IEC 15444-12

Then there are the other questions such as:

- Is the media being captured to removable media, or directly to a server and then file transfer?
- Is it getting encoded to a stream and transported over an IP link to the broadcast production center, where it will be captured as a file?
- What about stream format? Is it ASI or IP—is it J2000, MPEG2-TS, or MPEG4 H.264?
    - In the case of JPEG2000, it is defined as both a file and stream format.
- What file format is the broadcast center standard?
- What format is the original recording?
- What is the bitrate?

- How much compression should be done in the field?
- Which audio format is used? Compressed or uncompressed?
- How much storage is needed?

These are just some of the questions that need to be considered when looking at acquisition.

What factors are involved in deciding how much storage is needed? Here is one example: 1 hour of DVCPro100 creates a 60GB file. However, with 8–16 channels of audio, the file size is closer to 72GB.

There are a number of storage options when the essence is being captured on removable media. Optical disks are now four layers and can handle up to 128GB. Flash drives hold up to 128GB. Hard disks go into the terabyte range, with Solid State Drives (SSD) quickly replacing spinning disks. Speed is very important when making a decision on solid state and hard drives, since higher bitrates need faster drives. One of the major considerations when selecting storage is that bandwidth throughput and drive speed have as much of an impact on design as the *amount* of storage. This will be covered in more depth.

Standards and Formats

Articulating the difference between standards and formats makes for an interesting discussion. While the industry standards organizations have agreed on the standards and protocols, each lists them with their own catalogue numbers or the standard is listed as a joint effort:

- SDI is SMPTE and ITU
- MPEG is ISO and IEC
- JPEG2000 is ISO/IEC
- ON2 is Open Source

Once one group establishes a standard, the other groups may adopt it and assign its own designation. For example, the main MPEG2 standard is both ISO/IEC 13818-1 and ITU-T Rec. H.222.0. While these organizations publish them as standards, in reality they are considered protocols. This is something akin to a "polite" suggestion of a standard that allows many variations, formats, or "profiles." Whether it's a stream or file, there are many formats that are permissible under each standard. For each published standard, there are numerous profiles— and profiles are not necessarily compatible with each other.

MPEG Profiles

- MPEG2 4:2:2 @ MP/HL
- MPEG2 4:2:0 @ MP
- MPEG4 Part 2
- MPEG4 Part 10 (H.264)

In an SD/HD-SDI infrastructure, media transport is well defined. In the IP infra-structure and the transport of streams and files, ***bandwidth*** is king. When it comes to file standards and formats, there are many other decisions to be made. The basis for these decisions is determined by these questions:

- Where is the file going and how many resolutions are needed?
- Is it the library copy?
- Is it the production or proxy version?

This drives the decisions surrounding the following:

- The file size
- The amount of compression
- The format compatibility between systems
- How much bandwidth is available
- How much storage will be needed

Some file formats are used for encoding, while others are used for transport. When content is encoded, there is a separate codec for audio and video. When content needs to move across systems, it is wrapped into containers. All the codecs and containers are based on the published standards.

Containers:

- MXF – SMPTE 377M
- GXF – SMPTE 360M
- ProRes – Apple

When a stream is created for broadcast contribution, there typically are only a few standards used: MPEG2, MPEG4 H.264, and JPEG2000. These can also be encap-sulated into DVB S, DVB S2, DVB-T, or as ASI or IP.

There are other formats used for delivery to online and mobile platforms, such as Flash, ON2, HTTPLive, and HTML5.

When a file is being created, there are literally hundreds of codec options for encoding. Once the audio and video is encoded, it is then packaged into containers or wrappers. The codec may be AVCi. XDCAM, DVCPro, and ProRes for video and PCM, AAC, or MP3 for audio, but the wrapper is MXF, GXF, and QuickTime.

When making a decision on compression formats, is there really such a thing as lossless compression? With compression something is always given away! In the great codec wars, there are so many different formats and more coming every day. The codec decision is based on what the best format is for library, production, delivery, and proxy. The multiple delivery formats are specified for each of the distribution channels. While in production, interoperability is crucial if more than one edit system is going to be used. It is fairly typical to create the content in the multiple formats required for wireless, mobile, and web delivery.

Consider the subject of bitrates. All bitrates are not created equally; there are different formats and bitrates for files and streams. Quality is *not* only determined by the actual bitrate. There are other factors, such as the color space, bit depth, and inter-frame vs. intra-frame. Is it LongGOP? When choosing the bitrate, the higher the bitrate, the larger the file, which means more storage and more bandwidth. With a lower bitrate, there is a reduction in bandwidth and the ability to make more efficient use of storage. At the same time, you have to determine whether quality is compromised.

So what is the best choice? What's the best resolution to start with? One suggestion is to create one house format and look no further (a decision that can just as well be made in the tape-based world). Historically, it may have been 1" Tape, BetaSP, DigiBeta, or D5. There was one format for acquisition and one for in-house. Once a tape came in, a dub was made, and then the master was filed in the library.

In the IP- and file-based space, there are different copies of the file: a high resolution for the library copy, mezzanine resolution for production, and lower proxy resolution for searching and browsing.

Codecs
So Many to Choose from

Audio Video Standard (AVS); OpenAVS Blackbird FORscene video codec; Cineform; Cinepak; Dirac; Schrödinger; dirac-research; DV; Firebird [1] Original FORscene video codec; H.261; FFmpeg H.261; libavcodec); MPEG1 Part 2 (MPEG1 Video); Cinema Craft Encoder; FFmpeg; Ligos LSX MPEG1; MainConcept MPEG1; TMPGEnc; H.262/MPEG2 Part 2 (MPEG2 Video); Canopus ProCoder; Cinema Craft Encoder; Elecard MPEG2 Video Decoder; FFmpeg; InterVideo Video Decoder; Ligos LSX MPEG2; MainConcept MPEG2; TMPGEnc; H.263; FFmpeg H.263 (libavcodec); MPEG4 Part 2 (MPEG4 Advanced Simple Profile); 3ivx; DivX; FFmpeg MPEG4; HDX4; H.264/MPEG4 AVC or MPEG4 Part 10 (MPEG4 Advanced Video Coding), approved for Blu-ray; CoreAVC; MainConcept l; QuickTime H.264; Sorenson AVC Pro codec, Sorenson's new implementation; Vanguard Software Solutions; x264; Indeo 3/4/5; MJPEG; FFmpeg; Morgan Multimedia M-JPEG; Pegasus PICVideo M-JPEG; JPEG 2000 intra frame video codec; OMS Video; On2 Technologies TrueMotion VP3/VP4, VP5, VP6, VP7, VP8; under the name The Duck Corporation: TrueMotion S, TrueMotion 2; Pixlet; Apple ProRes 422; RealVideo; Snow Wavelet Codec; Sorenson Video, Sorenson Spark; Tarkin; Theora; FFmpeg; libtheora; VC-1 (SMPTE standard, subset of Windows Media Video); VC-3 SMPTE standard; Avid DNxHD; FFmpeg; Windows Media Video (WMV); WAX (Part of the Windows Media Series)..............and so on; and so on.

FIGURE 2-3

So Many Formats, So Many Choices

Not to stir up confusion, but indeed there are probably between 800 and 1,000 different codecs available. And it seems like more and more are introduced every day. Arguably, there may be only a half dozen or so that are mainstream and that matter, but there are many other formats out there, as well as something that is of greater concern: old formats don't die. When planning and designing a facility, it is strongly recommended to standardize on one format and maintain that throughout the entire facility and workflow. *THE GOLDEN RULE IS, "PICK ONE."*

Pick any one. Choose a bitrate, resolution, and file format compatible with production systems. Next, select the container, and then *STANDARDIZE ON IT*.

Then, communicate this standard to those creating content and keep it in that format while it's in the production environment. This way, while it is in-house, everybody's handling the same format.

Handling files and streams philosophically is no different than with tape. While it is important in the tape-based world to avoid creating a large number of copies to minimize degradation, in the IP- and file-based world, it is equally important to avoid multiple transcoding or encoding processes to minimize any loss of quality.

The same is true for library and production resolution. On a file transfer, files that are born digitally (e.g. XDCAM, P2, and HDV) are best left in the file format in which they were created. Taking a 25MB file and turning it into a 100MB file for archive, doesn't make it any better, and in many cases makes it worse. Keeping content at the original bitrate and format preserves the integrity.

How many formats are really necessary?

To begin, consider the internal formats for library (archive), production (mezzanine), and proxy (browse). These are the fundamental house masters. Then, there are delivery formats. These are different for each platform and delivery network.

An example of this would be On-Demand delivery, which requires a higher bitrate, and depending on which On-Demand provider, they have a format specification. For example, iTunes requires ProResHQ at a bitrate of 88–220Mb.

As far as streaming, the format depends upon the platform, so the format and bitrate that go to a phone, tablet, or PC will all be different. Moreover, when we consider mobile and what is available on broadband, the bitrate is going to be determined by the carrier.

There are a lot of format discussions and competition in the mobile market. While flash has been the dominant format for some time, HTML5 and HTTPlive are now becoming the new standards.

It's all in flux

MPEG4 H.264 has been the mainstay of mobile media, and H.265, MPEG7, and MPEG21 are on the horizon. The primary reason that formats are in flux is that

MPEGx x.xxx is a licensed technology, and the Open Source community is pushing other formats that are royalty-free. When it comes to choosing formats and compression, what looks good on a 65" TV is very different from what looks good on an 11" tablet, 3" phone, 1" wrist device or heads up glasses. The delivery platform plays a large role in this decision.

Ingest, Acquisition, and Capture

Table 2-4 compares the different format standards, their bitrates, and the average file size for an hour of content. As 4K becomes mainstream, it will have a considerable impact on production. One hour of 4K @ 3.82Gb/s is 1.72 TB/hr taking into account how much storage 4K uses, flash and optical become very limiting. Even at MPEG2 bitrates, the files are large and need a lot of bandwidth for transfers.

Format	Name	Bitrates	File Size	Platform
MPEG2 4:2:2) @ MP/HL	XDCAM	25, 35, 50Mbit/s	18 30 GB/ Hr.	Production
MPEG2 4:2:2)@ MP/HL	DVCPRO	50, 100 Mbit/s	30- 60GB/ Hr.	Production/Library
MPEG2 4:2:2)@ MP/HL	ProRes 422HQ/ DNxHD	147, 220 Mbit/s	100 GB/Hr.	iTunes/Production/ Library
MPEG2 4:2:0 @MP	DVCAM/Firewire	25 Mbit/s	15GB/Hr.	Production
JPEG2000	J2K	250Mbit/s	137.2GB/Hr.	Production
4K RAW	4K	3.82 Gbits/s	1.72 TB/Hr.	Production/Cinema
MPEG4	Blu-Ray	40Mbit/s	6GB/Hr.	DVD
MPEG4 Part 2	H.263	700K-3Mbits/s	N/A	Video Conference/ Web
MPEG4 Part 2	H. 263	700K-3Mbits/s	300MB	You Tube
MPEG4 Part 10	H.264	700K-3Mbits/s	300MB	Web, Mobile, Flash
MPEG4	H.264/AVCHD	100Mbits/s	16GB/Hr.	Production
HTML5-Open Source	H. 264	700Kbits/s-3Mbits/s		Apple iPhone, iPad,
HTML5-Open Source	Web/VP8	700K-3Mbits/s		Web, Mobile

FIGURE 2-4

The table shows how big an MPEG2 file can be for an hour's worth of media based on different encoding formats and bitrates. It's easy to see that with MPEG4 there's a substantial difference. One thing that is not well publicized is that actual file size will vary based on the audio encoding. Each HD-SDI stream can have eight (8) pairs of embedded AES, plus the ancillary data channels. One example is, an XDCAMHD 50Mb/s that claims 30GB for an hour with uncompressed audio is really 62Mb/s, so the total file size is really 37.5GB. Using JPEG2000 at 250Mbit/s and adding audio and data brings the true bitrate to 318Mbit/s.

The transition from analog to digital maintained a similar architecture of video and audio routing and distribution. In the analog world, there was tape-based A/V record and playback, serial control (RS422 and 232), and IP was in the province of the IT department.

As tape evolved to SDI, the audio is now embedded in the video, and digital signals have different requirements for routing and distribution, such as re-clocking. With the introduction of workstation-based editing (NLE), IP appeared as a control layer in addition to serial protocols.

Tape machines have evolved into servers, and IP has changed from being only a control layer to fully encompassing media and management. Before, there were separate paths and layers for audio, video, and control, with a completely separate ecosystem for management. Now everything is fully integrated—audio, video, control, and data for management all travel in the same IP stream or file. The stream is transported and received by a variety of applications and then parsed for craft, distribution, and management.

FIGURE 2-5A

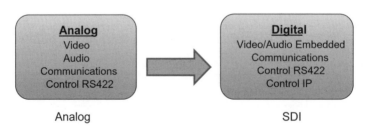

FIGURE 2-5B

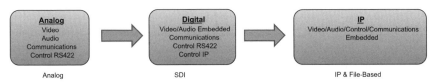

FIGURE 2-5C

Ingest

Let's take a look at Ingest in the file-based architecture.

There are different ways to encode an IP file. Studio cameras output HD-SDI, 4K, and—coming soon—8K and 16K. HD-SDI enters the router, production switcher, and/or goes directly to a production server. Once there, it is captured and encoded to a full-resolution file while a proxy is being created. Both are placed in storage. All these processes are handled by the media manager. The higher resolution 4K, 8K, and 16K use multiple outputs at the camera and need to be recompiled and synchronized at the router, production switcher, and encoder. These higher resolution formats have different transport and handling requirements, and the higher bitrates impact editing and storage.

If there are lot of Isolated (ISO) camera shots, then a larger number of ingest channels are needed. How many concurrent encoding processes can one server handle? As the master file at high resolution is being created how many proxy versions are needed and do they need to be created simultaneously? This will impact application servers, network traffic, and storage design considerations. The impact on storage will be in throughput and bandwidth, not so much actual storage capacity. The design consideration becomes how many concurrent read/writes the storage architecture can support. How many can the disks handle at the same time before requiring additional disks (spindles) just to support the load? Solid state storage has the same issue.

Case Studies

Sometimes the number of concurrent operations are not based solely on the number of camera ISOs but on concurrent events. Here are two different case studies showing the challenges and considerations when there are a number of concurrent events.

Case Study 1

A large global humanitarian organization modernized its meeting facilities. There are seventeen (17) meeting rooms that operate at the same time with three (3) to five (5) hour meetings in each room every day. Each meeting and therefore each video channel has multiple audio channels for different languages embedded. There are two (2) high resolution recordings of each meeting plus three (3) different proxies for each recording, with automated and manually entered metadata to log the meeting events. The encoding processes start/stop are triggered automatically from a scheduling system sending the information as an XML file. There are over 100 real-time editing processes on the files as they are being written. Separate processes take the live content and encode it to the web and mobile media, carrying the language channels and inserting captioning.

In this example there are multiple simultaneous encoding processes creating multiple bitrate files for each instance of encoding. There are multiple streams transported to storage where files are being written as multiple users are using proxies on a timeline to create subclips and rendering (processing) new files that add to the writing demand of the storage and the management demand of the application software. At the same time metadata is added and associated correctly to each file. This is one demonstration of scale and the amount of concurrent processing, reading, writing, and network traffic of large files that can present a considerable challenge in designing the architecture to support this business requirement of the organization.

Case Study 2

Most American sports leagues are centralizing their replay review process to provide greater integrity and accuracy to the officiating of the games. To accomplish this, American sports leagues have instituted centralized game archiving and instant replay review systems. They are transporting all the cameras live from every venue to a centralized production center, encoding them individually to file for archive and production while having real-time access for replay review of an infraction.

One such example is having as many as nine (9) cameras from over twenty five (25) venues individually transmitted into encoders at as many as five (5) bitrates for archive, production, management, and distribution. There may be as many as nine (9) concurrent live games on any one (1) day. This translates into eighty-one (81) ingest channels simultaneously encoding and producing four hundred and five (405) files concurrently. All while any or all of these files are being accessed by production and official review in real-time. Once a replay has started, another file is created only for the replay segment with all nine (9) cameras in sync. This translates into nine (9) x five (5) files being read and controlled with the same metadata to track plus the unique information for each file and then writing forty five (45) separate files, all while the master file for each of the nine (9) cameras continues to be encoded and written.

While all this is going on, there is a steady stream of statistics and tournament data automatically being ingested in addition to manual metadata being entered in real-time. All this metadata has to be assigned correctly to each master, proxy, and sub-file (parent-child inherited metadata properties).

This level of ingest puts an incredible burden on all the systems. The storage has to perform multiple read/writes on the same files as new files are created while the master files continue to be written. The applications that manage this move files and streams, plus in separate layers perform processes that enable a user to display and work with the files. Other applications or middleware are performing automated processes to integrate data or trigger other processes with metadata. All while management applications are indexing and registering the media and metadata for searching and browsing. Now add in quality control with management of all the processes, traffic, and bandwidth. It is easy to see how the systems can be stressed and why the planning and design of the entire architecture is so critical.

Studio

Studio production has a different set of requirements whether it is live or recorded. Studio production allows for greater control and management of the processes. Many if not all of the encoding and format decisions have already been made.

When media is contributed from the field as SDI, it arrives in the facility and is encoded to a file in the house format and bitrate. If it's an IP stream it will be encoded in the field, streamed to the broadcast center, and captured to a file as it's received.

In the IP architecture, SDI video and analog/AES audio are still part of the core in studio production. Cameras output SDI while microphones and speakers are analog. Once the output of a microphone enters the audio production console, it becomes AES. Another possibility is when the microphone is wireless, the receiver encodes the analog to AES prior to connection to the production console.

One of the reasons that the output of cameras and live audio remain as SDI and AES, when compared with the transition to IP- and file-based production for editing, graphics, distribution, is that the real-time visual transitions (i.e. crossfades, wipes, and keying) and live audio mixing with processing still heavily rely on SDI and AES technology. The industry is reluctant to adopt IP for *live* production. IP has not overcome the hurdle of the seamless transition between SDI sources, or AES audio that has to be mixed, equalized, and processed, cross fades or the type of on-air keying, layering, and effects that wideband—SDI and AES—production devices do. Some of the same challenges are present in 4K and 8K production.

Another major change in the evolution to IP is in the command and control systems. This has brought significant benefit and change to studio automation. Automation is so much more than master control program origination, however it did become critical in multi-channel and multi-platform program origination. Automation is also a major component in the studio production control room. There are different levels of studio automation, from using automation controllers that manage production equipment to complete "Studio in a Box" systems and everything in between. And all of these are managed and controlled over an IP architecture.

In the studio, there are robotic camera pedestals and mounts that integrate with virtual sets controlled by dynamic touch panel graphic devices and image displays that tie into the virtual set using telemetry integrated with the camera mount and lenses. This is all managed in the production control room.

The camera robotics are controlled over the IP network and using an application running on a server that manages and controls the camera motion with the ability to memorize all the parameters of the shot that includes motion, pedestal position, pan, tilt, zoom, and focus. All of these can be pre-programmed and embedded in a script. When the script runs it automatically recalls the metadata and triggers commands or it can be manually triggered by an operator using a touch panel.

In a fully automated control room, the system operates the switcher, mixer, triggers pre-programmed effects, triggers and inserts graphics, and plays program clips. The graphics and clips are managed with metadata that has embedded control information. In this scenario, all the command and control data is over IP, with the automation system as an application on a server based on a database. The graphics, animations, and packaged program clips are sent to the play-out server that runs under the automation system for production. The integrated production program is streamed to the delivery server for capture to file as well as sent as a live stream directly to distribution and will also be converted to many different formats for each of the distribution platforms.

What is the "Studio in a Box"?

The "Studio in a Box" concept is a server-based system that has all the capability of a production studio integrated in a suite of software applications—virtual audio mixers, video switchers, still store, digital effects, clip players, and computer graphics. Most of these offer a physical control surface that while it is still operating a software application, the control surface has the look, the feel, and acts like a dedicated audio mixer and video switcher. These "Studio in a Box" systems are all in the IP infrastructure, and while they are all server-based, they are technically hybrid since camera and audio mixing is handled as AES/SDI.

In a "Studio in a Box," the production can be highly automated. Thus enabling the director/producer to build the script with a shot list for the cameras including the positioning for robotics, and create a clip play list that incorporates the graphics, integrates program elements like transitions, switches between the live cameras, and creates a complete studio production program with very few people. Today's studio operations include robotic camera operations based on server applications, encoder control and management, metadata tagging and logging.

The director/producer can create templates that will allow for pre-programmed production scenarios.

There are also downsides to a fully automated "Studio in a Box," such as the fact that the on-air production scenario cannot change dynamically. If an on-air presenter speaks out of turn or is out of place, they are off camera and potentially off mic. The cameras and microphones track the script and blocking.

This changes the design of the control room and studio. Going back to a studio that's not in a box. The control room and studio are still more efficient to manage and operate. The control room monitoring uses multi-viewers that are flexible and integrated with the router. These multi-viewer displays include source identification, clock, countdown timers, audio levels, and tally that are all fully integrated using metadata for onscreen information. There is a lot more information available to them and it's all easily accessible. Production switchers now include sophisticated multi-channel visual effects, store and play still images, and have the ability to play animations and program clips. The production switcher is integrated with the router and reads the router table for input labeling using metadata. The audio console connects to a server that has all the inputs and outputs on a frame. There can be many more inputs and outputs that the console can support in any one session, however it can also function like a router or interface to core router and populate input channels based on a specific production requirement. The operators can store a show's profile on a removable drive so as the studio is used for different productions, they can easily restore the configuration for any show. All of this is controlled over networked devices that communicate using the command and control metadata.

Automation and Metadata

Within the IP architecture and production workflow, files need to be accessible to various applications and processes and possibly need to move between storage locations. All of these processes are applications on servers and where the file may never move from its location for an application to perform a process or have a user access it, there are also processes that require the file to transfer between application servers and storage locations. There are different layers of automation to manage these processes and handle the file movement.

Metadata, which is discussed in detail in a later chapter, plays a crucial role in all aspects of the IP- and file-based architecture. In studio production, metadata is the control layer integrating all the devices, it triggers the automation systems, it manages the Studio in Box, and provides the tracking information for the camera robotics and virtual sets telemetry. The control or structural metadata from the automation system will be associated with the stream or file and includes all the content elements, control activities (e.g. camera switching), as well as the administrative metadata, usage rights, permission, tracking, and descriptive metadata.

Craft Production

Craft production or editing no longer has to wait for the recording or "ingest" of a production to complete. In the file-based environment, as the master file is being captured and ingested craft editing can begin. All the systems are networked together and can access the storage. Editing can use a proxy and timeline reference that does not require the file to be complete and fully rendered. As the editor starts working, they may save their work in progress in a separate work area in the storage environment. Sometimes this is completely separate storage, however, it is still connected to the core storage architecture. There are areas sometimes referred to as "scratch bins" and these store interim files before the full program is completed. This enables an editor to move between systems and access their work. The edit systems can also access content from the media library and archive using browsing tools. This is critical in news and sports. While a live event is still in progress and being recorded, a news or sports editor can be creating a story or trailer clip and sending it to air while the live event continues.

Where graphic production is not typically real-time to an event, the graphics team still can access any content as it is being ingested. Additionally the graphic workstations can access the media library for any content and then place the finished product as files for inclusion in a production or as a clip for play-out on the play-out storage. One of the imperative changes in workflow is that each time a file is created or edited new metadata needs to be added.

Removable Media

Removable media was first used to refer to analog tape and later to digital tape. Remote production still records to digital tape, but now also records to optical disk, flash, solid state storage, camera attached hard drives, and portable digital video recorders (DVRs). These portable DVRs are either hard drives or flash, sometimes with onboard controls or controlled via a computer interface.

Removable media is the primary method of capturing in the field. While ENG/SNG typically have had attached storage onboard the camera, sports and news remotes that use big production vehicles that have large servers and vast amounts of storage. Some have removable hard drives that once the event is over are put into a suitcase and hand carried back to the broadcast center. *Therefore, the volume or capacity of the removable media becomes a critical decision.* In making this decision about which is the best type of removable media to use for a specific production there are several considerations:

- What is the availability of a specific removable media type? Can it be found in the field or is it necessary to bring enough?
- Can decisions be made in the field that would allow content to be deleted and disk space recovered?
- The ease of use—is it hot swappable? Does anything need to be formatted or reset?
- How transportable is it? Will it fit in a pocket or backpack, or does it need a hardened suitcase?
- Capacity: how much content can it capture?
- How much additional storage do I need to bring with me?
- Will it immediately be transferred to other storage?
- What is the cost of the media per hour of content?

When it comes to removable media, reliability and durability are big concerns. When recordable media was tape-based, there were issues of physical damage, humidity, delamination, tearing, or getting caught in the machine. While DLT or LTO tapes are still susceptible to the same issues, they are typically used in a controlled environment. Optical disks are pretty resilient, however they can't withstand being run over by a car or getting scratched, but other than that appear to be sturdy. Hard disks have moving parts and need an interface (USB, IEEE1394, Thunderbolt, Ethernet, etc.) and since hard disks are still magnetic media, it is important to consider how they will be transported and handled. Will these disks remain viable indefinitely? This will likely require a migration strategy (see Storage and Media Management).

The introduction of Solid State Drives (SSD) as a replacement for hard disks solves some of these issues, however, present some of their own. There are concerns about their durability. How long will they last? What's their sensitivity? Are they fragile? Solid State Drives require specialized card readers, and these are frequently ungraded and changing. The continuing evolution of these technologies creates compatibility challenges, so it is reasonable to question their backwards compatibility. There are many considerations to be taken into account for even the simplest of storage decisions.

Streaming Contribution

In addition to capturing content to files and transporting the media back to the broadcast center, the output of a camera or production switcher is SDI, UltraHD, 4K, and 8K (multiple HD-SDI outputs) that is transported as a live stream directly to the broadcast center where it can be integrated for delivery and distribution while being ingested and captured to a file for media management and archive. Contribution from remotes back to the studio has evolved from a dedicated path satellite, fiber, or microwave feed. There are new mesh or fabric networks using Fast Ethernet, Synchronous Optical Networking (SONET), and Multiprotocol Label Switching (MPLS) that provide private network high bandwidth transport enabling full resolution (uncompressed) or compressed content to be streamed. Many of the broadcast occasional service interconnect providers (Encompass, The Switch, Vyvx, etc.) are providing new tiers of managed services over these new mesh networks. Where in the past their services provided transport of a single SDI signal per fiber, they are using multiplexing technology that can send multiple signals on fewer fibers. There are many other technologies used for contribution such as open Internet, VSAT, BeGan Sat Phone, bonded 4G modems, cellphone, smartphones, and Skype. While these tend to transmit at lower bitrates and lower quality, receiving content from them has become more than acceptable if the story is interesting enough or there is no other way of getting the remote feed back to the broadcast center.

The open Internet presents a number of issues, there is no guarantee of consistent bandwidth from the origination point to the receive location; the second issue is security; many countries have policies that block moving video over the Internet and Internet transmission typically has considerable network latency.

Another alternative available for contribution is cloud services, which is finding its way into file and stream delivery for media. A cloud provider relies on private network internally and either private network or open Internet for the first and last mile. Some cloud services are used for simple file transport, however there are some service providers that provide optimized bandwidth products in the cloud that improve the speed of transfer. Internet-based cloud services have the same challenges of bandwidth limitations as direct transmission over the open Internet. If the cloud service uses private networks like Fast Ethernet or MPLS then it is more reliable and more secure. While the cloud is not completely private, it is also not the fully open Internet.

The cloud is more practical for file transfers than live streams. There are a few services that are starting to offer live delivery in the cloud including encoding. This becomes more distribution than contribution. There are also production services that are cloud based (this will be covered in a later chapter). Using the cloud for live streaming will have a few drawbacks. Bandwidth is a major issue, and if it is over open Internet then there are network latency issues that may impact delivery. If the content being delivered is time sensitive, i.e. live sports, then latency is a problem.

The cloud is still dependent on the famous "*last mile*," meaning that reasonable bandwidth in the field is required to transfer or stream the captured media in a timely fashion. Remember bitrates! The amount of bandwidth determines the speed of transfer and the size of the file or the bitrate of the stream will determine the deliver time.

In remote production and live contribution there have been changes, instead of using the remote truck as a production control room, the cameras and microphones are encoded directly to IP, multiplexed, and sent to the broadcast center. There, they are decoded and brought into a production control room, which does the switching and mixing. Communications and return monitoring is sent back over IP to the venue. In this scenario, IP enables multi-camera production to be controlled from the broadcast center. The broadcast center now uses its own studio control room without needing a production vehicle in the field. It is easy to see how latency can affect production in this scenario.

Case Study

Beginning in 1996 for the Atlanta Olympics, NBC made an interesting decision. Instead of building an extensive and expensive master control and commercial integration facility on site, they had a significant amount of fiber connectivity established from Atlanta to their Rockefeller Center New York City broadcast center. They built their International Broadcast Center to handle production locally, a high percentage of editing and graphics were produced in NY and then transferred to Atlanta for insertion to production. Then the fully integrated program was sent back to NY for commercial integration and broadcast.

In each subsequent Olympics, they added more capability, over time they were able to change the workflow and production on site by using high bandwidth connections and transferring content to their broadcast and production facilities in the US. Using their media management infrastructure, all their production teams were able to access the files for production and distribution. Where the bandwidth allows, the camera feeds and audio are sent to their US production facilities and the program is switched and produced there.

three

Setting up new workflows and processes is the next piece of the puzzle, and this chapter looks at these changes. The changes in technology in the move to an IP architecture are not the only ones and it requires many other considerations.

This chapter focuses on the new and tighter integration between business and production operations, which includes workflows and processes within the business organization. One of the biggest challenges in the transition to IP- and file-based technology is change. The adoption of new technologies that replace long standing technology solutions coupled with the resultant workflow and process change is intimidating. Many manual processes are now performed by automation. The processes and workflows that do require manual support are different.

Adding to that is the introduction of a new type of governance. Governance is a principle that has been in the IP world for a long time. While the media industry does have rules and policies, it was not structured and characterized as governance. Governance is the rules and policies that are critical to the smooth functioning of the business. All of these elements are critical when planning or evolving to the IP- and file-based architecture and must be tightly integrated. In the IP ecosystem, governance is the business rules and policies that are programmed as part of the configuration of the applications that manage and control the movement of media.

- Which databases can be integrated?
- What fields are protected and not accessible by a group of users, another application or system?
- Do these databases need middleware to communicate with each other?

These are only some of the decisions to be made before any application configuration can be done.

Workflows and Business Process

The word *workflow* is much used and abused in the broadcast industry, where it is used to describe almost everything. The definition of workflow is "a sequence of connected steps":

- *Work* is defined as a depiction of a sequence of operations by a person or a group.
- *Flow* refers to content, a document, media, or a product being transferred from one step to another.

When applying it to handling IP- and file-based media, *workflow* is the process of an individual, a group, a software application, or a dedicated device that is doing something with content in sequential steps as it moves through a series of processes within the ecosystem. These operations are handled by both automation systems and human interaction.

The core to most media workflows can be quantified by three (3) things:

People: These are the champions, sponsors, stakeholders, and users that rely on the workflow to do their job or manage the business.
Process: This is the stewardship that guides and supports the successful execution of the workflow.
Technology: These are the devices, applications, and systems which facilitate and enable the workflow. Each of these is a critical element when changing the workflow from tape-based to file-based, from manual to automated, and the conversion to an IP architecture.

Integration between Broadcast and Enterprise

In the IP ecosystem, production and business processes, operations, networks, and systems are more tightly integrated and use automated processes to manage the media workflow. Business systems such as legal, finance, marketing, rights management, and business intelligence all create and need access to metadata.

Metadata is the foundation of the IP architecture, and each business department contributes metadata that controls media movement and manages access. All users within the organization will search and browse the library for different reasons and operations. There are a number of touch points between the business enterprise and the broadcast network for managing media traffic. The business units need to establish working policies that allow the permissible data to move between systems across networks, while the IT and broadcast IT groups need to work closely together to create seamless workflows for the users.

In the IP- and file-based ecosystem it is harder to segregate how the technology and workflows are used throughout the entire ecosystem and architecture. There is such tight integration that in describing the overall technology architecture and design requirements, it is difficult not to constantly refer to where workflow fits.

Not only do the technology systems integrate differently in the IP architecture, the workflows integrate differently as well. And how they integrate is very important. Figure 3-1 shows that media management is one of the core processes in the IP workflow.

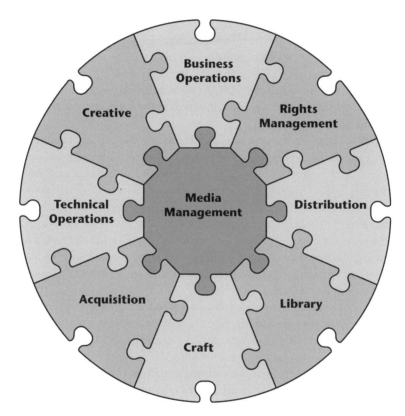

FIGURE 3-1

This demonstrates that while each of the business and production processes are interdependent, they all integrate to the media management system. There is a mix of media and business departments that each can add permissions, rights control, management metadata along with the descriptive and automation control metadata. More importantly, the IP ecosystem brings new operational and production processes; and new workflows.

Media Management – This is the heart of the environment and ecosystem. It encompasses all the technologies, the complete workflow and set of business processes that manage the movement of media from the time of its creation throughout its entire lifecycle. Each cog in this wheel are the workflows and processes that represent the core groups that are accessed directly or indirectly by a user or are involved in media management.

Creative – This is the beginning of the media lifecycle. All other processes stem from the creative process. The descriptive and administrative information developed in the creative process is the first set of metadata entered into the media management system and could be entered prior to any media being actually produced. This may include the script, technical resources, locations, contracts, usage rights, and other early production information.

Acquisition – Once production begins, titles, location information, media logging details, production notes, contribution information, and technical details are entered. The acquisition process includes the ingest and encoding process and, of course, requires media handling. During the acquisition process additional control and management metadata is entered, as well as the application of retention rules set by the business groups.

Craft – These are the processes that transform the content into the finished asset. This can be accomplished in a studio control room, with editing and/or graphics operations. This step encompasses all of the different craft operations that provide the finishing details before the asset moves to distribution. In the craft process, any content that's created or acquired and will be included as an element to the program or separate asset and need metadata to become associated with main asset under a set of 'parent–child' policies in the media management architecture. The element (child) inherits the properties of the main asset (parent) while retaining its own identity. In the IP environment each asset can be both a parent and a child.

Library – In the IP architecture and workflow, the digital library serves a similar and more expansive role than the tape librarian or archivist. This is the command and control center for metadata and media management. The librarian oversees ingest and file movement processes, ensuring the integrity of the metadata. Managing the media archive is another of the workflows the library manages. In the file-based ecosystem, media in the library is available instantly for production, however, once it's archived the retention policies take effect.

Business Operations – The legal, finance, rights, marketing, and business intelligence departments are the primary business groups that are directly connected to the media lifecycle and ecosystem. These departments contribute and extract metadata for contracts, sales, marketing, rights management, clearance, and market analysis.

Rights Management – Rights management has a larger purview than just usage and copyright protection. User access is also *an internal* management issue (e.g. setting permissions to access content). It's not uncommon for certain assets to need encryption or protection from specific users for production or distribution reasons. Rights management includes content expiration rules and consumer access control. Rights management metadata controls what delivery platforms have access to the content.

Technical Operations – Technical operations include engineering and encompass all of the file handling and movement throughout all the systems and processes of the IP infrastructure. While most of the content handling operations are automated, there still needs to be management and quality control. With all the different delivery platforms, technical operations include insuring that the various complements of metadata as required by numerous distribution networks are included and correct. Quality control is part of technical operations and is responsible for assuring the integrity of the media from the moment it is ingested to media movement, file management, and play-out. All of the program origination center functions are typically considered part of technical operations (e.g. error logging and discrepancy reporting).

Distribution – Distribution first changed from single channel to multi-channel origination. The next change was to add in multi-platform delivery and multi-channel origination. Each distribution platform has a different format, requires different metadata, and has different rights and protection rules. Program origination is typically fully automated.

The workflows and business processes come down to the relationship between people, process, and technology as it applies to all the stakeholders and to the core processes that are integrated within the file-based and IP technology architecture.

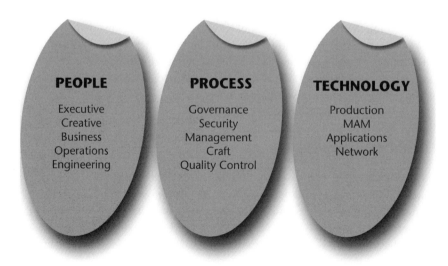

FIGURE 3-2

People – These are the stakeholders and executives that want access to the media by using search and browse functions for presentations, creative ideas, and promotion. The creative teams begin the media lifecycle by creating the asset through the production and craft processes. The operation units control and manage the media across the entire ecosystem, while the business units establish the governance and monetization policies. The overall responsibilities of engineering include ensuring that all systems are operating properly at all times, protecting operations from any interruptions in services and providing quality control for the files and streams as they move throughout the infrastructure.

Process – The area of process integration must cater to a new alignment of the primary workflows and processes that support business, technical, and production operations. Governance is the rules and policies that manage, control, and protect the asset though its entire lifecycle. One of the critical processes is security, protecting the asset, and controlling access. This protects the value of the asset and manages the integrity of the systems within the infrastructure. The management process handles file movement, metadata, library, and distribution. Craft production is no longer an autonomous operation from the rest of the media lifecycle. Content is registered and indexed before being accessible by craft production. The craft process includes updating metadata and authorizing the content to move within the media management workflow to library and distribution. Quality control is one of the most critical processes. This monitors and insures the integrity of the file or stream for library and distribution. With the number of formats and file types, maintaining quality control is key to monetizing the asset.

Technology – This is a good place to remember that this is really all about technology; Applications, Servers, Network, and Storage. This is looking at the technologies and all the other requirements that need consideration in the design and planning of a facility. The technology is the support systems that enable the new processes, workflows platforms to exist. There are new technologies for production, MAM is a common place acronym and applications and network are part of our normal lexicon. The IP- and file-based media technologies are the core foundation to the new lifecycle of media.

Business Integration

Business integration is the newest component in the broadcast and production workflow. The IP ecosystem opens up a new opportunity to integrate the business technologies and workflows responsible for monetizing the assets with the actual media workflows. These groups previously worked in operational and technology silos with each system isolated from the others and interconnectivity limited if at all.

Business operations is comprised of a group of departments that, in the IP architecture, is integrated across the network and allows applications and databases to communicate and share data governed with rules and policies to control access. Let's take a closer look to see where the integration occurs and which design considerations come into play on the application and network side required to make this work.

The legal department can interface directly with the entire media process, beginning at the creative process where production contracts are created and managed and then extending all the way to delivery, where distribution contracts with restrictions and rights management are imperative. The legal department needs access to the media management system to input management metadata and export metadata into the applications that support the legal processes.

The finance group retrieves records from the media management system in order to track costs and resource utilization. One of the roles finance performs is to assign monetary value and track resource costs associated with media creation and production. The finance applications provide the monetization rules and policies to the media manager as metadata. This metadata will stay with the content as it travels to each of the delivery platforms. When a consumer attempts to access the content by either paid subscription or conditional access, this is controlled by user authentication and validation rules and policies that are managed by the finance department.

The marketing group is another major stakeholder. From the outset of production, marketing has data that tracks and stays with the content as it travels through

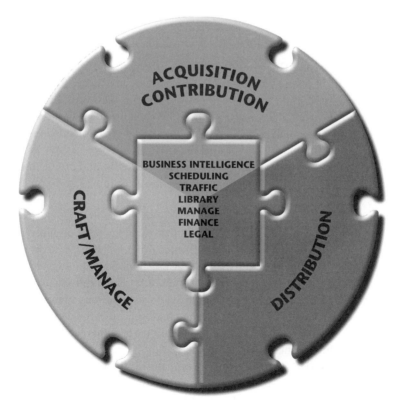

FIGURE 3-3

its lifecycle. Marketing creates promotional products that need to be assigned as "children" of the main asset. This is done in the media manager by adding associated metadata. The metadata that travels with the asset to the different delivery platforms and independent program guide services (data aggregators) is also created by the marketing group.

Marketing manages the tracking data embedded in the media (e.g. watermarks and encryption keys). The entire value chain of media in the multi-platform world broadens out to include metadata for clearance, subscription access, tracking, plus provides the necessary information to Electronic Program Guides (EPG) and Program and System Information Protocol (PSIP) (for Over-the-Air). One of the most important applications of metadata is when it is sent to the delivery platforms to enable indexing and search by all the different recommendation engines and devices in media enabled devices, players, and DVR/PVRs.

Bringing this back to the design of the IP architecture, the marketing applications need to place this metadata into the media management system and have it travel between applications and across networks as an element of the asset's metadata. The accessibility of media to the consumer is fully dependent on the metadata—as is the monetization of the asset.

Business Intelligence

Business Intelligence (BI) grew out of the requirement and ability to gather data from all types of online services and the different social networks. It is the process of retrieving data from various content distribution platforms and bringing it back into the business for analysis. There are specialized application and database tools that can analyze the data to create a set of recommendations that can be used for critical business decisions. Business Intelligence has moved into the media industry and become a valuable tool for marketing and production. There is a considerable volume of data that can be gathered or mined from content users across all the platforms.

Business Intelligence is defined as the process of seeking information and knowledge for the purpose of making quality decisions that are in line with organizational and business goals. It is having factual information, analytics, and statistics to do predictive modeling based on historical facts that are available immediately (in seconds instead of hours, days or weeks). This is the metadata mined by all the delivery platforms and returned to the broadcaster for analysis. It can be an enormous amount of data, in which case it is referred to as Big Data. Business Intelligence uses this data to gain powerful insights, obtain knowledge, and use it to predict the patterns of the end user.

Business Intelligence is all about the quality of decisions made within the organization AND can be a primary competitive differentiator. It is designed to support the decision-making process and facilitate informed actions to meet business goals. Leveraging analytics and the knowledge gained from it makes business intelligence an essential new weapon for competitive media companies.

Business Intelligence also refers to the software applications and techniques that are used to identify, extract, and analyze data. This is data such as consumer information, search results, and appearances in recommendation engines. These statistics and data sets are a result of the metadata that is sent to the Electronic Program Guide (EPG) aggregators, other program guide service providers, and by tracking the interaction with end users.

Business integration has implications across the workflow and business processes and greatly depends on technology. Business Intelligence relies on large databases with algorithms that analyze the data and is used to develop rules and policies for other data sets that manage the content. Database integration is at the core of media management.

Operations Workflows

The Production and Technical Operations groups need to adopt new workflows and processes driven by the integration of the IP technologies and infrastructure. These are some of the groups that need to adapt to the IP- and file-based technologies. This will have a large impact on how they achieve their program creation and origination requirements.

In the production workflow, as the media is captured in file form or encoded to a stream, there needs to be a process to assure that the media is properly indexed, logged, and tagged. Some of this is handled by automated processes moving data between systems while some still requires a person watching and entering metadata as a production or event is taking place.

During live sports, there is a considerable amount of data that is created prior to an event. The team rosters or list of competitors, their numbers, statistics about them. Once the tournament starts both automated and manual metatags are entered. Car racing has many sensors in the cars and around the track. In other sports where embedding sensors on a person or in their clothing is impractical, a logging person needs to watch the event with the ability to enter tags with more than a timestamp. The logging systems use "Hot-Keys," touch panels with custom interface specific to that sport or dedicated controllers. Each of these interfaces have preprogrammed buttons with embedded metadata so when the logger touches the button or key the tag has the type of the event, possibly the player number, court position, and other data. This allows editing systems to automatically retrieve clips and elements for production.

Another of the efficiencies in file-based production is that during the encoding process while the media is being indexed, logged, and tagged, craft editing begins simultaneously using metadata to assemble clips. This provides a huge benefit to live production such as sports and news. Within the craft production process, producers, directors, or non-technical managers can browse the digital library and select content. The identified elements can be assembled as a list or even into an edit decision list (EDL) and sent as direction to an editor to conform or include in a finished program. These can all be concurrent processes; however, each time a different user accesses the content for inclusion, there should be an entry made in the metadata record. There are new rules and policies to govern usage. Governance is a larger subject that is covered in greater detail with media management.

In the production workflow, the ingest or editing process needs to allow time for rendering to finish a program for delivery and archive. During ingest, rendering is the process that takes the temporary file created and finalizes it for use in production or delivery. In the editing process, once all the decisions are made using timeline-based editing, rendering is the process that assembles the elements into a single finished file for delivery and archive. When rendering during an edit session, the original file is untouched and a new file is created with its own metadata, also inheriting metadata from the original content. Rendering is similar to what was called a conform in linear editing that took an Edit Decision List and created the final program for distribution. Rendering is necessary with file-based systems, because for editing they use proxy or mezzanine resolutions and then need to render the finished program drawing the elements from the original high-resolution content into a new file which is the finished program. For ingest, typically a temporary file is created until the ingest is complete, and then the file is rendered to enter the media management system. The rendering process can be both slower and faster than real-time based on the parameters of the finished file.

There are many changes in the production workflows. Post-production is integrated on the media network to local and shared storage that is managed by the library. There are changes in the edit process as to how programs are finished. The editing workstations are craft devices and there are dedicated servers to render the final program without burdening the edit workstation, thereby freeing it to continue working. The format differences for content on the various platforms change the editing and post-production process. Editing for web, tablet, and phone is different from a large screen display. Each delivery platform has its own production requirements.

Technical operations is all of the processes and workflows that manage media movement throughout the infrastructure from ingest to distribution and is responsible for quality assurance. Technical operations include transmission, master control, traffic, library, archive, and engineering. It is responsible for the integrity of signal quality and the quality of all content in archive. The transition to IP- and file-based technologies has had a significant impact on technical operations and to most of these processes. One major change is that many of the processes are handled by automation.

In the IP ecosystem and value chain, the digital library functions as the primary media manager, this system owns the governance and retention policies that move media between different storage locations and controls access. The library is responsible for confirming the integrity of the metadata, tracking missing data, and validating it so that the asset has all the appropriate metadata it needs for archiving, distribution, and browsing. Technical operations handles format conversion for both ingest and delivery. And, of course, quality control is key. Quality control has different processes depending on if the media is a file or a stream.

Technical operations has an expanded dimension of workflow in the IP eco-system, providing users with remote access. This encompasses a variety of user groups; they can be managers, production personnel, editors, media loggers, exec-utives, and many others. The design of the IP infrastructure has to account for these users, not only in the physical and technical design but in the operational workflow design as well. Looking at the different workflows across production, technical, and business operations involved in media handling and management, many of these can be done using remote access.

Network management is core to the IP infrastructure and there are workflows associated with the configuration of the network to support technical operations.

Figure 3-4 shows how the main network segments (enterprise LAN and broadcast LAN) are segregated yet connected. A firewall is used to protect each segment and the firewall enables and controls protected Internet access to both segments. On the enterprise is legal, finance, marketing accessing application servers, databases, and file storage. The broadcast LAN has media, command and control, media management, communications, production accessing application servers, data-bases, and storage. The systems on the enterprise LAN and broadcast LAN need controlled access to each other.

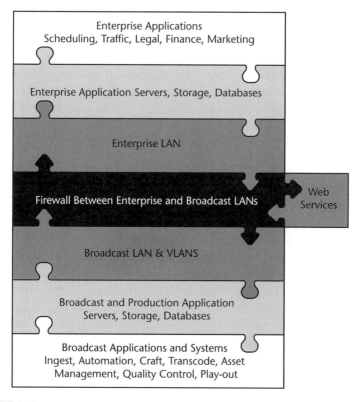

FIGURE 3-4

The search and browse engines of the media manager make media and meta-data easily available using remote access and this enables a user to assemble an EDL and leave only the rendering to be processed in the main infrastructure by automation. In this scenario, an edit can be initiated and mostly executed remotely. The media can be easily accessed using search functionality for all operations via the browse mode.

Media management is based on rules set in all IP-based systems. This too can be managed through a remote interface. Setting up this access is a joint operation between the IT and Broadcast Engineering groups, and is based on access rules and policies under the overall governance schema.

Automation

In the IP- and file-based ecosystem, automation is used to manage and move the media and metadata as it moves between servers, applications, and processes.

Introducing this level of automation changes the workflows at each production and technical operational process. Automation is more than just play-out; auto-mation manages the movement of media throughout the infrastructure. Systems monitor other systems waiting for media to arrive. Once the media appears in a folder or as a stream, automated processes recognize its arrival and trigger the next event. This could be an encode or transcode, or it could alert a producer or editor that there is media available for editing. The digital library is alerted that new media is in the system and to check metadata. The quality control tools monitor incoming content locations and begin testing the integrity of the media and metadata.

Moving media requires rules and policies and systems to manage them. In the IT enterprise industry these applications are called Business Process Managers and they control the movement of files, the integration of databases, and the interac-tion between applications sharing data. In the broadcast IP infrastructure these sys-tems are called conductors or orchestration applications. One of the key differences between enterprise and broadcast file management is that in the enterprise docu-ments and data do not move throughout different processes and applications. A word document stays with word processing, a spreadsheet is not converted back and forth between applications and data sets that may be shared by multiple applications don't change in structure. In the broadcast architecture, media and metadata moves and changes as it travels between different applications and processes. There are file movers and data directors. This is where orchestration becomes a critical piece of the puzzle. As the media enters the system, it needs guidance to go from acquisition to ingest, get encoded, get registered to the management system, possibly transcoded, and converted before being moved to the appropriate storage location.

Orchestration is automation on steroids. Each system needs to know what to do, when to do it, and what to do with the file after it is done with it. Systems have profiles for performing their tasks. Different files have different priorities and need to be tracked as they move through the entire media management system out to distribution. What if there is a problem with the file or a process, something needs to be aware that there is an error and alert an operator or engineer. What about resolving conflicts when there are a number of concurrent processes all happening requiring the same resources?

As the media moves through its lifecycle, traffic information is automatically pushed into the system, directing the media to all the play-out destinations with the associated metadata. The traffic system needs to manage and produce formats and schedules for each platform and distribution channel.

We took a brief look at studio automation and some of the changes driven by IP-based technologies. Now, let's look a little more closely at these technologies and how automation changes the studio production workflow. Both the studio and studio control room have changed based on IP technology. In the studio, camera robotic control is all IP, the pedestal and pan/tilt head receive command and control data of the IP network. LED lighting instruments have integrated dimmers receiving command and control from the server-based dimmer control over the IP network. The camera lens sends and receives data for the robotic telemetry. The studio set may have a large image display that is server controlled showing video and graphics or the studio can be all green screen and use a virtual set controlled by automation running on a server. The virtual set allows the real camera to move in the 3D space, while the graphic image from the virtual set is rendered in real-time from the same perspective. This virtual scene adapts to all the camera settings (i.e. pan, tilt, zoom, track) and it supports multi-camera production.

Most if not all of production devices are servers with their control surfaces connected over an Ethernet network. Most if not all of the devices in the studio and studio control room have an IP management and control layer. Setting up for a production involves accessing the server to configure the production system. The entire control room can have a "soft configuration" which essentially means the operator can recall a "profile" from a database or flash drive at the control surface to set up the control room for the specific production. The production system may use a server-based browser or application to recall the configuration.

The introduction of automated systems into studio production has defined new workflows. There can be a robotic camera operator or the TD (technical director) can use a touch panel interface pre-programmed with camera positions and moves. In live studio production like news, the camera position and shots are pre-programmed based on the scripts for the program. This includes recalling and playing clips, inserting graphics and visual effects, and running the teleprompter, all from a unified control screen that is either a touch panel or by using a keyboard and mouse. It's not uncommon to have a single operator in a control room overseeing an automation system running the show.

There are new workflows in the post-production area. The introduction of Non Linear Editing (NLE) technology completely changed the workflow of post-production. Now, with all these systems integrated over a network with shared storage, file movement and media management play a larger role from when the editing system was a closed architecture. The producer and editor use the media management tools to locate content, and content is instantly accessible during the ingest process. From a workflow perspective, using the media manager a producer can create a storyboard complete with an EDL. Depending on the complexity of the finished program, automated processes can use the EDL and assemble a finished program. Or an editor can recall it and finish the program adding new elements and transitions while at the same time entering more metadata. Adding metadata makes finding and using the content again easier.

Production workflows and processes are continuing to evolve and the next generation of these processes will be cloud based. There are a number of editing and graphic systems being offered as cloud services and how these will integrate into the post-production workflow and processes is also evolving. Does the content live in the cloud or are the proxies in the cloud making the content accessible from anywhere and the EDL is created in the cloud while the full resolution content stays in the broadcast center storage? Once the EDL is created in the cloud, is there an automated function that retrieves it and pushes it to the rendering engine to create the final program?

Traffic Workflow

Taking a step back and looking at the entire lifecycle again, the content travels through the infrastructure towards distribution. It's time to explore the changes the networked and integrated environment has created with traffic and scheduling. Since traffic and scheduling have been using databases for a long time, the first evolution was from the connection between traffic and master control as a device-to-device RS232 serial connection to an Ethernet connection that transfers data over a network to many devices simultaneously. Traffic systems have continued to evolve to having metadata communicating between devices and systems while being managed by automation.

Multi-channel distribution was one of the first big changes in program origination workflows. In the first generation of multi-channel origination, the delivery format was the same to each platform (cable, satellite, and DTV) and different programs were delivered via different channels. Now in addition to the same format for multiple programs that are delivered for delivery to multiple channels, there are many different formats that are delivered to the many different delivery platforms and there are multiple transmission paths that also are in many different formats.

This has significantly changed the workflow and processes of traffic operations. For commercial networks the integration of commercials is substantially different for each platform. Linear program channels have one type of commercial integration, On-Demand has a different one and streaming to devices even another. There are differences streaming over broadband vs. to mobile. The traffic system has to manage and control the delivery to each of these platforms, and each platform is different: different in what it receives, how it receives it, and how the asset is formatted specifically for each distribution channel. The scheduling group has to produce multiple schedules. There is one schedule for linear distribution for more traditional viewing, and a different one for On-Demand delivery to all devices and DVRs that are programmed by show rather than linear channel. Programs are watched based on the viewer's interest and a user-created schedule regardless whether the origination is still linear.

These transmission and distribution processes are all managed and controlled via automation. The traffic and scheduling systems create the metadata that is sent over the network in a common data format and are instructions to the automation system.

The traffic department manages playlists not only for multiple channels and time zones, but also across multiple platforms. The traffic logs being fed to the automation systems have to include the instructions that specify which metadata fields accompany the content to each platform. There is no standard for metadata for the different content distribution channels. The traffic and scheduling process provides the command and control metadata that are the instructions to the automation system for what type of processing the content requires for distribution. The scheduling department needs to manage these instructions.

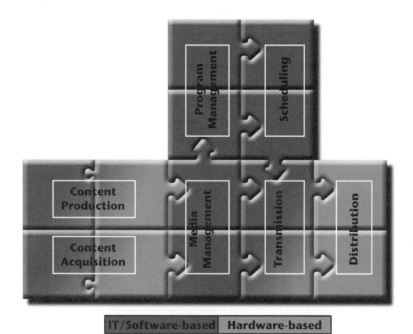

FIGURE 3-5

Roles and Responsibilities Redefined

The roles and responsibilities of support personnel within the organization change as part of the transition to an IP architecture and workflow. First there are expanded responsibilities and a need for new skills and knowledge base. Then there are new positions created as a result of IP stream- and file-based processes. The ingest manager is responsible for bringing media and metadata into the system. This can include file transfers, encoding, and transcoding. The ingest point is when metadata tagging and media logging begins. In the file-based ecosystem, the digital librarian ensures that the asset management system logs the media and creates the proxies, and that the metadata is entered properly for users to be able to search and browse.

The "master control" operators now manage delivery to multiple platforms that have their own specifications for media format, resolution, and metadata requirements.

Engineering has expanded responsibilities that add additional workflows and processes in the IP ecosystem. Problem solving, maintenance, and support for the IP infrastructure are mostly software centric. Even the hardware systems are maintained with software tools. Routine maintenance on software systems includes updates and service patches that occur on a frequent basis. Maintaining hardware includes firmware updates, storage, and bandwidth management. The maintenance process for servers, storage, and network or proprietary IP devices is not pulling it out of a rack and putting it on a test bench. That may apply only if a part needs replacing and not even then. There is less bench work in the IP- and file-based ecosystem. Keeping track of software and firmware versions plus device configurations is as much a part of engineering as designing, building, and maintaining in the more traditional sense.

The entire architecture and infrastructure resembles an IT datacenter and an IT backbone. The broadcast engineer needs to be skilled in network routing, switching, servers, applications, and storage. Understanding the interfaces between different software systems and middleware is an essential knowledge base.

There are many pieces to the puzzle that make up the complete IP- and file-based architecture. Understanding how all the business units work together is just one more piece of this complicated puzzle. There are new workflows and processes as a result of the technologies that are integrated within the IP architecture. There is a need for media and metadata to be accessible across all the business units that rely on database integration.

Workflow and process are an important design consideration. User access and the interfaces between software applications are all based on the way business processes interact.

four

Media management is probably the most important piece to this puzzle. In the IP- and file-based ecosystem, media management is the core of how technology, operations, workflow, and business processes all interact and function. AND, at the heart of media management is metadata. The logging and entering of metadata is essential because it includes the critical information used to package, sell, and distribute a show. Metadata is associated with the media at all stages of the media management lifecycle.

Creating and managing assets in the IP- and file-based ecosystem is dependent on using descriptive and administrative metadata that stays with the asset as it travels through the value chain and into distribution. It is imperative to have well-organized and well-structured metadata associated with every asset. This metadata is what content delivery networks use to enable program guides, recommendation engines, and search tools to locate the asset, with the metadata serving double time for business intelligence requirements. In the file-based ecosystem, it is critical to have well-defined policies that are rigorously and consistently adhered to, thus enabling the creation of quality metadata that is correct and complete. Without this, deploying the latest in Digital Asset Management (DAM) technology becomes an empty vessel.

Media management encompasses the management and movement of media throughout the infrastructure, from ingest out to delivery and distribution while providing metadata for search. Media management includes tracking viewer access and managing the return data from the distribution networks on usage statistics.

On the business side, media management and metadata are central to the monetization of file- and stream-based media.

Even before the first element of media is created, there is metadata, administrative and descriptive metadata. As media moves through the entire system, more metadata is continually added, and the metadata also controls how it moves and where it goes.

In the file-based lifecycle, there needs to be a management structure, complete with standards and policies to govern the media flow.

Media management encompasses more than asset management and metadata. Indeed, it is the entire movement of media within the IP- and file-based ecosystem and value chain. Media management is an overarching term that describes the handling and control of streams and files across the media lifecycle and value chain.

Media management is not a new concept; however, in the IP- and file-based ecosystem, it has become the core and foundation of the architecture. From the moment a production is planned, there is valuable information—data—that follows the idea into creation. Once the essence is created, if it doesn't have something describing it, identifying it, and managing it, it has no value. Metadata defines the essence and turns it into content that, when associated with a program that will generate revenue, becomes an asset.

Figure 4-1 works from the top down, it shows how media management functions are the core processes as content moves from acquisition through creative and out to distribution.

Media management starts with business drivers and user groups. This drives the policies that create the workflow and processes. This is the entire basis for the metadata that travels through applications and databases, that runs on servers and dedicated devices over the network where the content ultimately lives in storage and archive.

All of these elements are part of the media management system that ensures the media flows through the infrastructure:

- It creates the relationships between assets, contracts and rights
- It controls the movement and processing of media with automation
- It manages the library and retention policies
- It organizes media for search and browsing
- It provides the structure for the organization and cataloging of media
- It manages the formatting and delivery to distribution platforms
- It delivers program data to distribution channels for guides and searching
- It imposes the usage and access rights to the asset
- It analyzes data aggregated from the various distribution networks
- It governs the entire media flow with standards, rules, and policies

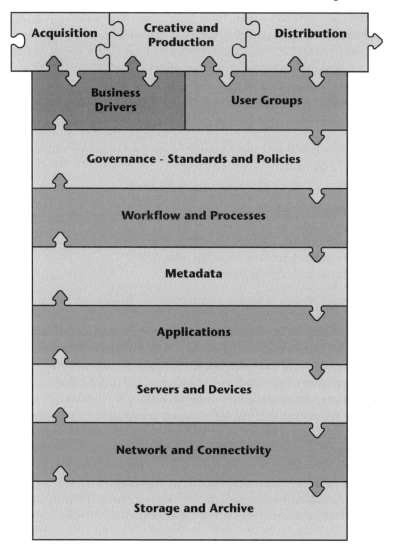

FIGURE 4-1

Metadata

The term "metadata" has become almost a generic term almost like using Kleenex to describe facial tissue. What is metadata? Metadata is often called "data about data" or "information about information." In telecommunications metadata is the personal information on a mobile device, metadata is the embedded key terms that search and recommendation engines use to find content. In broadcast, metadata takes on the responsibility of managing and controlling every aspect of media from before it's created to when it goes to archive forever!

In broadcast, metadata is structured information that describes, explains, or makes it easier to find, use, control, or manage digital stills, video, or audio.

The word "metadata" was originally coined in the library world, and is now commonly used for any formal scheme of resource description, applying to any type of object, digital or non-digital.

The following are some of the ways metadata is used, as well as a few terms used to describe metadata:

Digital ID—This is a unique identifier of the asset or object, which could be part of a file name.

Discovery—Metadata allows resources to be found by:

- Defining multiple criteria
- Identifying resources
- Bringing similar resources together
- Distinguishing between dissimilar resources
- Giving location information

Organize—Metadata is used to index, organize, and catalog media based on taxonomies and ontologies.

Interoperability—This is the ability to have multiple systems operating different hardware and software platforms, data structures, and interfaces to exchange data with minimal loss. Assets can be searched across the network more seamlessly by using well-defined metadata schemes and shared transfer protocols.

Automation—Automation uses metadata to provide instructions and communicate between different systems for moving files throughout the ecosystem, controlling applications and devices that manage and process media.

Library and Preservation—Metadata typically centers on the discovery of recently created assets. There is always the need to retain programs, B-Rolls, and elements. However, there is also concern that digital assets are fragile and they can be corrupted or altered, intentionally or unintentionally. The assets can become unusable as storage media, hardware, and software technologies change.

Metadata is the key to ensuring that assets will survive and continue to be accessible into the future. The archiving and preservation processes require special metadata elements to track the lineage of a digital object.

Metadata not unlike "regular" data is managed in a database. The field structure of the database and the relationship of the fields are commonly called the "metadata schema."

When a metadata schema is being planned, the discussion usually centers on the tags and indexing that will assist in finding an asset. For retention, the key question is, "What needs to be preserved for the future?" Metadata is much broader than that.

There are six main types of metadata:

- Descriptive
- Structural

- Administrative
- Rights Management
- Preservation
- Technical

Descriptive metadata is the most familiar. This is the metadata used to browse and search for an asset. It can include elements such as title, abstracts, descriptions, author, and keywords. Most of the catalog and organizational tools are used with the descriptive and preservation metadata. This is the metadata that search engines and program guides use to locate media.

Structural metadata is the control layer that handles the movement of media, automation, and interaction between applications and databases. An example of structural metadata is recognizing a file in one format, sending it to a transcoder, and giving the transcoder a set of instructions to convert the file to a different format using a specific profile. Once the file is converted, it triggers a QC process to validate the file. This process controls how compound objects are put together or which instructions are sent to transcoders to format for cable, tablet, and web.

Administrative metadata is the management layer—defining the media by file type, ownership, location in the storage network, internal usage, and access rights. Additionally, this is the metadata that integrates with business systems. Essentially, the administrative metadata captures the technical details of the asset, making it easier to manage.

Rights management metadata is different than administrative usage permission and access rights. This is the protection layer of distribution, and includes copyright, intellectual property, and distribution rights. It carries expiration rules, encryption policies, and tracking metadata used in watermarking for piracy protection.

Library & preservation metadata is used for archiving purposes. When metadata is part of any discussion, it typically centers on archiving. In the IP- and file-based architecture, the digital library takes on greater roles and responsibilities. The library retains and manages the "master copy" of all content ingested to the system. There are core elements that need to travel with the asset for preservation, ensuring that when it is accessed or restored in the future, there is adequate information (both descriptive and administrative) that identifies it.

Technical metadata describes the creation and technical characteristics of digital and physical objects, such as file format, resolution, size, duration, and track configuration. The automated capture of detailed technical metadata is central to obsolescence planning and preservation strategy.

Creating, managing, and integrating metadata is at the core of a well-planned and designed IP architecture. These functions must exist from the origin of media during the creative process all the way through to production, acquisition, craft production, distribution, and monetization.

Media management relies on the integrity and consistency of metadata. The amount of metadata expands as the content travels across the architecture, with each system and operation adding more.

Media management is not a single application, it is a collection of integrated applications. There are a number of different technologies and systems within the IP architecture that manage and handle media. The metadata moves with the media. The systems and devices need to be able to receive and communicate information about the file or stream.

Many of these systems use different database applications and structures. There needs to be middleware, a connector application that facilitates interoperability. It functions almost like a traffic controller and translator. There are a number of these types of applications in the marketplace. They operate on a rules-basis that knows which data is permissible for extraction or loading to each of the applications and databases that are part of the media management system.

Media and metadata move throughout the entire media management infrastructure. It is a critical requirement that there is interoperability between devices and systems. It is not uncommon for different systems to use different databases and database structures, which means that there needs to be controlled integration among databases. In the IP ecosystem, the term "governance" is used to define the rules and policies that protect access to sensitive data while enabling integration.

"Interoperability" is one of those words that instills fear in the hearts of all system engineers. System and device interoperability is a myth that has been passed around the broadcast and IT industry for many years. First, there were little black boxes and software widgets to deal with device interface. Then, as IP- and file-based technologies arrived, so did even more devices, APIs, SDKs, and middleware to deal with "interoperability." Published standards do not resolve this!

Media management doesn't solve it, either; it only adds another layer. In the IT industry, interoperability keeps programmers employed writing interfaces and creating special mapping or custom translators to move data between applications and databases.

Media management is all about encoding, transcoding, and transmuxing in order to allow the media to move "seamlessly" among systems and devices. Metadata faces the same challenges. While the eXtensible Markup Language (XML) has become one of the accepted standards as a file transfer protocol for interoperability, the format and field structure of the XML is not part of the standard. The formats CAN be different in each system that uses metadata, and depending on the role of the device and/or system, the metadata fields WILL be different.

Another consideration is how the XML is delivered:

- **Synchronous**—The systems need to communicate dynamically in both directions, ensuring that the metadata is the same in both systems.
- **Push**—One system sends the XML file to another system programmed to receive it.
- **Pull**—One system has a folder or IP address configured, and under a schedule or trigger, it retrieves the metadata and updates itself.

When a metadata interchange happens between databases with dissimilar field structures, the metadata has to be re-mapped to match the structure of the receiving database. This is sometimes done manually, with custom code or by using middleware specially designed to interface disparate databases. In the IT community, there is a process called ETL, or "Extract-Transfer-Load." These are middleware applications that can be considerable problem solvers in the media architecture across databases.

These tools operate under a set of policies programmed into the application that control the access to only the metadata that is allowable for transfer, thus creating a seamless transfer.

In the media management system, there a number of applications, dedicated devices, databases, and storage that are organized and structured in a multi-segmented layered network. The metadata files moving in an XML format travel across these networks.

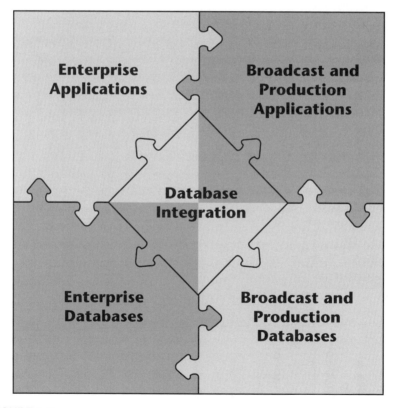

FIGURE 4-2

Figure 4-2 oversimplifies the number of layers within the network architecture, but it clearly demonstrates that there exists an application layer, database layer, and storage layer.

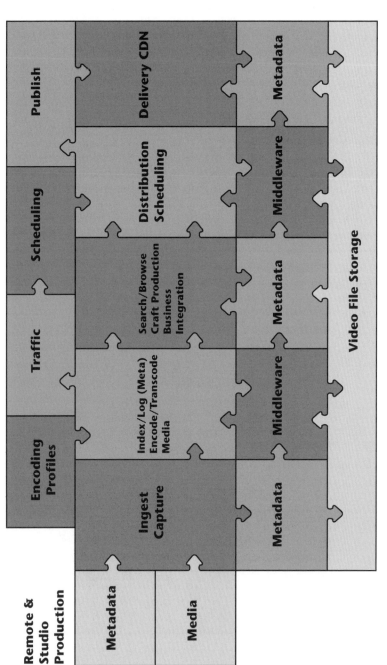

FIGURE 4-3

The figure also shows the different applications that potentially share metadata across databases while the media is organized in a federated storage architecture— possibly by department, workflow (ingest, production, distribution), or simply application.

There is a management layer to support the interoperability between applications and media. One of the critical design considerations (and one that demands constant monitoring) is the traffic and performance requirements of each of the applications and systems. Metadata and media move across the network differently and to various locations. The media management systems control and manage this movement, locating and integrating the appropriate metadata to each of the applications and delivering the media to the next stop in the lifecycle. At the same time, it maintains the relationship of the media and its associated metadata.

There is a dynamic interaction of XML files moving the metadata, sometimes requiring middleware to enable this.

"Integration" is a broad term that boils down to which systems really need to interact, have access to each other and why. Here are some of the business systems needed to interact with media management:

- Traffic and Scheduling
- Automation and Control
- Marketing and Business Intelligence
- Executive Management
- Legal and Finance

Traffic and scheduling are somewhat obvious. The traffic system will handle multiple schedules going to different platforms, so it is easy to see how the delivery metadata plays a key role. The traffic system sends a control set of XML along with the traffic information to the play-out server, and gives an instruction set to the automation system. The scheduling system has the electronic program guide (EPG) and search metadata that may be different for each platform and must be packaged with the asset for distribution.

The automation system controls file and stream movement throughout the infrastructure and into the different storage areas for access by each of the different production and business systems. It receives its instructions in XML from the traffic system, finance, marketing, and legal.

Marketing has a number of roles and responsibilities. It needs access to the content to create promos and ads, and it adds metadata for tags and indexes for search and recommendation engines. Business intelligence compares the return metadata with the original metadata sent to the distribution channels and provides direction for the tags and keywords that marketing includes.

There are a number of user groups that will only use the search and browse features. Executive management will research assets for many different reasons.

Legal and business affairs require access to oversee usage and access rights. They will enter contractual metadata that controls which internal and external users can access, handle, and use the asset. The finance group accesses the metadata to see what resources have been used to create and manage the asset. Finance may also add metadata of other associated costs not captured by other systems.

Metadata Standards

Just as there are media standards, there are also metadata standards:

- Media Formats – MXF, GXF, ProRes, DVC, AVCi, DNx
- Metadata Schemas – Dublin Core, EBUCore, PBSCore, ISO/IEC, SMPTE
- Data Interoperability Formats – XML, ODBC, CSV, J2E

In the previous discussion on codecs and format standards, it was important to ensure the container for audio, video, and metadata conformed to an interoperable standard since this is essential for delivery across multiple platforms. The same is true about metadata. With metadata there are schemas; this is the field and data record structure in the database that will host the metadata. For a title, there is the structure of the title, it will be considered a text field, it will be allowed to be "X" characters long, have fields for categories and have certain properties and characteristics for indexing.

The Dublin Core metadata schema refers to a 1995 Invitational Metadata Workshop hosted in Dublin, Ohio, by the Online Computer Library Center (OCLC), a library consortium, and the National Center for Supercomputing Applications (NCSA). The "Core" refers to metadata terms as "broad and generic, being usable for describing a wide range of resources." The Dublin Core Metadata Initiative (DCMI) is an open forum for the development of interoperable online metadata standards.

During that fateful conference, the Dublin Core established an agreed standard of the essential fields an asset must have for a librarian to manage it. Metadata needs a format to move between systems, and XML is the current preferred format.

No different than the media industry, in the library industry there are a number of organizations that create and manage the standards and protocols for metadata.

These organizations have established standards that clearly define where metadata lives within the asset, how it is transmitted, and what metadata the essence must have that is placed directly into it by a camera or encoder.

When it comes to standards and protocols, the metadata community is not to be left out and it most definitely has its own set. There are a number of standards and protocols associated with metadata:

- NISA
  - METS
  - MODS
- OCLC/NCSA
  - Dublin Core DCMI
- SMPTE
  - SMPTE 335M-2001 Metadata Dictionary

- ANSI /ISO
  - ISO/IEC 11179

The National Information Standards Association (NISA) has a set of guidelines for the use of metadata, this is mostly for library and information management, however it can include non-digital materials.

- METS is the Metadata Encoding and Transmission Standard developed by a group of research libraries that is widely accepted.
- MODS is the NISA Metadata Object Schema.

Most archivists will agree that the Dublin Core is the minimum requirement of metadata fields and details for any digital object that is to be preserved.

As metadata migrated into the broadcast industry, SMPTE, ISO, and ANSI jumped in and created a set of metadata standards specifically for the broadcast industry, as well as the transport protocols for how the data set is associated with the media file (XML).

There are new standards for the layer of metadata embedded with a file. MPEG7 and MPEG21 specifically address metadata and the second screen platform delivery.

The SMPTE Metadata Dictionary (SMPTE 335M-2001) is composed of a set of elements describing audio/video that can be grouped into categories:

- Identification
- Administration
- Interpretation
- Parametric
- Process
- Relational
- Spatio-temporal
- Organizationally Registered Metadata
- Experimental Metadata

The ANSI-ISO/IEC standard consists of six parts:

Part 1 - Framework
Part 2 - Classification
Part 3 - Registry meta-model and basic attributes
Part 4 - Formulation of data definitions
Part 5 - Naming and identification principles
Part 6 - Registration

An interesting note is that the ISO/IEC 11179 standard does not describe data as it is actually stored. The two main purposes for this standard are:

1) Definition of the data and
2) Exchanging data

While it appears that there are many slightly differing standards, they all refer to a controlled vocabulary and dictionary that carries the descriptive, structural, and administrative metadata. All metadata is organized in this structure.

The example below shows the transition from raw essence to monetized asset through a simple example using water as the essence and then applying metadata to turn it into a valued asset.

Data for Water

- **Essence:** [**type**]Spring Water
- **Metadata:** [**object**] H_2O: [type] spring water: [location] Vermont: [date of origin] 20 March
- **Content:** [**category**] A bottle of water
- **Asset:** [**title**] Vermont Water {branded for sale}

The water itself is the essence; by adding metadata descriptors, [object] H_2O, [type] spring water, [location] Vermont, and [Date of Origin] 20 March, it becomes categorized content. Once it is branded, it has value and can be monetized.

The object and category are descriptive; the type, title, location, and date are administrative within the organization of metadata.

Now apply the same concept to media. Here is an example of how metadata turns media essence into an asset.

Data for Media

- **Essence:** [**type**] Audio, Video, Image
- **Metadata:** [**object**] HDD: [format] ProRes: [date acquired] 20 March: [location] Vermont: [air date] 20 June: [platform] mobile
- **Content:** [**category**] News, Sports, Entertainment
- **Asset:** [**title**] Emmy Winner Show

The essence "type" will be sound, video, or image. Adding metadata for the media format, date acquired, and location is what turns the essence into content. Next, a category is added. Finally, adding a title assigns value and completes the asset creation process.

Storytelling and the Value of Metadata

We will now discuss asset management, since metadata has been defined and it's clear what its value is in the handling and management of media. Up until now, the discussion has revolved around media management as a set of integrated systems and processes. It's now time to look at the value of asset management. There are a number of names used to describe asset management, "MAM" for Media Asset Management or "DAM" for Digital Asset Management.

This is not to be confused with DMS which is Document Management System which handles documents, PDF, PowerPoint, and spreadsheets or CMS which is a Content Management System that is a program that allows publishing, editing, and modifying content typically for web or mobile delivery.

Back to MAM/DAM. The value of asset management for media has increased substantially as the amount of media being consumed on different types of screens and platforms increases.

Metadata has not been the most exciting or interesting aspect of the transition to file-based media production. It has been treated more like a necessary evil, but all that has changed or may be changing. Metadata has always been valued as an efficient means of automating and managing file-based assets from production through scheduling and play-out. With broadband and mobile platforms, it has extended its usefulness to outside the broadcast center as an integral part of new ways that users can interact with media and is being used to open up new revenue opportunities.

This is the storytelling part. The most obvious example is live sports where statistical data is integrated with video. Metadata is not just the keywords embedded for search engines. It is companion information or an interactive component that encourages engagement. This is an essential element in social communities and boosts the stickiness of websites. Metadata is the Electronic Program Guide (EPG) that is used to display schedules and also used by recommendation engines during a search. Descriptive metadata empowers the viewer to find content using a keyword search.

There is a fair amount of attention focused on the volume and sophistication of information supplied for each program (movie, music, book, etc.), which can range from the technical data that marks a program as part of a series and assigns digital rights for distribution. It will show program duration and have additional editorial detail like still images, biographies, reviews, or recommendations.

There is another side to storytelling. The return data from the distribution platforms and social networks provides a wealth of data to be analyzed. Business intelligence, once gathered and analyzed, will tell a story about the viewers to the marketing department. There is an enormous amount of data that flows back to the broadcast center, challenging the IT network and overwhelming databases.

What is asset management? At its core, it is the indexing and cataloging of all content within the file-based ecosystem and once it is considered valuable enough to retain and archive, making it accessible under controlled access. There are many other features and functions within the asset management application that are part of the entire IP- and file-based workflow, but the primary function is to manage the media.

One of the main uses of an asset manager is to locate the asset. This is where controlled vocabularies and structured metadata are necessary in order for the search tools to perform properly and is key to having the correct vocabulary.

Librarians, archivists, and preservationists that manage large collections are all familiar with the terms "taxonomy" and "ontology." While the genesis of these terms originated in science and metaphysics, they are now very common when discussing asset management.

Taxonomy – Greek *taxis* "arrangement" and *nomia* "method" is the science of identifying and naming species and arranging them into a classification.

Taxonomy is used to create classifications according to a pre-determined system or controlled vocabulary. It is used with the resulting catalog as a framework for retrieval.

In asset management, taxonomies are used to organize assets and manage metadata. By employing a taxonomy to classify content and assets, it makes searching or browsing using a digital asset management tool easier for users who do not know many details about what they are looking for.

Ontology – Greek *onto-* "being; that which is" and *-logia* "science, study, theory" is the philosophical study of the nature of being, existence, or reality as such. It is also the basic categories of being and their relations.

Ontology is a classification scheme. It is a way of defining the relationships between objects in the world and organizing objects by subject categories. Ontology also defines how to divide up an object. This might not necessarily be by subject; an object may be divided instead by type, format, and location.

The following statement might be a touch contentious: "Asset Management" at its core is a library function. Asset Management is a process, workflow, and technologies. Looking at it from a technology perspective, it is the tool that catalogs, indexes, and makes media searchable and retrievable. There are different ways to search, such as by browsing the audio and video or by asking questions in a structured or unstructured way using keywords or pre-determined search criteria (filters).

The asset management tool must maintain associations between the finished program and the elements used to create it. There needs to be sufficient information for an archive when it is retrieved in the future.

The following are the primary functions and operations of an asset management system:

- Catalog
- Associate
- Archive
- Browse
- Retrieve
- Monetize

The asset management system is a complex software application based on a database, media players, and with multiple interfaces to connect to other databases and systems for automation, usage rights, analytics, and others. The application organizes the media according to the metadata that's been entered.

This assumes that media logging and tagging based on manual and automated metadata entries were actually made. Getting users to enter metadata consistently and reliably is a considerable challenge. The media asset manager organizes the media and enables it to be searched, accessed, retrieved, and annotated.

The asset manager catalogs the media and creates the associations using agreed and accepted metadata schema and rules (e.g. parent–child), with search tools that make access browsing easier. The asset manager controls the movement of media within the storage architecture between applications from ingest to craft to delivery and archive, packaging the metadata that will travel with the media to the distribution networks. This metadata has the program guides, interactive services, and information that will support additional revenue opportunities (i.e. promotions, specials).

So, now that Pandora's Box is open, let's take a look at Asset Management and the role it plays in media management systems.

The media/digital asset management system is essentially the digital library and provides a number of important operations in the file-based environment.

- Capture/Digitization
- Standardize/Transform
- Asset Management
- Production/Distribution
- Storage – Media & Metadata
- Quality Assurance

First and foremost, the asset manager has to recognize all possible formats in a multitude of resolutions and bitrates. It must capture the metadata in a robust database that is well organized so that the asset is accessible. It has to provide access to the media and metadata for browsing, preparation, and transfer to the production and distribution systems.

Many asset management products include a number other features and functions, such as automation, that are part of the entire media management system.

The asset manager and digital library is one of the core systems in the master IP- and file-based architecture and ecosystem. Beginning at the point of ingest, the asset manager tracks the media and metadata, logs them into the system and places them in the appropriate storage location. During the capture or ingest process, it manages the encoding and any transcoding to conform the media to the house format. It does all this while tracking and cataloging any processes and changes that may be occurring at the same time.

As other elements are used in the production process, the asset manager maintains the relationship and tracks the associated metadata. As the metadata is updated and expanded, it manages where the asset is and controls access and usage. When the media is transferred to distribution, the asset manager determines which metadata moves with the media.

The digital library manages all the storage locations in which media resides and tracks the movement between systems. Similar to a "conventional" library, the asset manager "checks media in and out" of the systems as it moves from ingest to production with archive and distribution.

The storage architecture is the foundation of the digital library. Storage can be segregated by location, process, resolution, or retention rules. This can be based on different types of media, high and low resolutions, and types of access (such as online, near-line, and offline), as well as access for other processes (e.g. ingest, production, distribution).

The digital library is the primary handler and controller of file movement. It must also own quality control as well. As files and streams pass through the system, the tools that analyze the files and streams need to process the media and, if there are issues, alert the engineer and attempt to solve the problem.

The asset manager possesses the search and browse functions that enable users to access and retrieve media using metadata. As previously discussed, there are metadata standards to facilitate this.

It all started with the Dublin Core, which established that a small and fundamental group of text elements would provide enough metadata components so that most media resources could be described and cataloged by using only fifteen (15) base text fields. When the European Broadcast Union (EBU) looked at establishing a metadata standard, they used the Dublin Core as their base and then created EBUCore which essentially added forty-five (45) more fields for a total of sixty (60) base fields now text, numeric, and alphanumeric. Then as the US Public Broadcasting System (PBS) moved to a file-based system they looked at the EBUCore and embellished on it to create PBSCore. For the most part, as standards go, it stopped here and now these are mostly used as a guideline or a base by everyone else in creating their own metadata schema and calling it "My OwnCore." While designing a metadata schema is always done with good intention and the best practices, most organizations create their own metadata schema with field sets that end up well into the hundreds. The US Library of Congress started with over 800 fields for metadata and who knows where it is today.

The asset manager maintains the metadata in a structured and indexed database based on taxonomy and ontology rules to organize the metadata. This structure becomes important as soon as a stakeholder wants to find something, then the search and browse tools come into play.

There are many different methods and types of search tools, it is interesting to note that most of them use one or both of the most popular techniques and algorithms that are known as faceted and contextual searching.

A faceted search is a structured technique for accessing a collection of information about an asset created by using a faceted classification. A faceted classification allows an asset to be assigned multiple sets of attributes. By assigning a large number of attributes to an asset, the classification can be ordered in multiple ways, allowing users to explore by filtering the available information.

Once an asset is categorized using multiple attributes, it can also be retrieved using multiple attributes. A user could use a single term or link together multiple terms which increases the chances of retrieving the asset they are looking for. For example, the search for an asset could be performed by having all the attributes filtered by date, location, format, and type of programming or by specific keywords.

A contextual search is more of an unstructured method, allowing the user to use random descriptors and letting the search engine look through all the metadata in an unstructured way, finding the most relevant matches. This type of search is more time intensive, requiring more CPU processing power. In the contextual search, it opens the search to more assumptions when returning results or enables targeted results based on the analysis of the search request. Contextual searching needs algorithms that can take random attributes or use metadata in an attempt to understand what is being searched for.

An example here is how advertisers can target users based on a search that has only a few parameters and then using recommendation engines, deliver a broad number of results promoting various products.

Searching through content and getting reliable results requires a good set of tools and integrity in the quality of metadata. However, in the absence of a good system, there are always more tried and true methods of searching like using a crystal ball, a magic wand, or pixie dust.

Storage Architecture

While the asset manager is the core application in the media management system, the assets need some place to live. The storage architecture is where the media and metadata reside.

There are multiple components to the storage architecture. Storage can be a local disk in a server, an externally attached storage device with one type of protocols (i.e. USB, Thunderbolt, and IEEE1394) and Network Attached Storage (NAS) or a Storage Array Network (SAN) with a different set of protocols. This introduces another layer as SAN and NAS systems will use Fiber Channel, SCSI, iSCSI, and Ultrawide SCSI as networks protocol to move files between disk arrays and also have an Ethernet interface for the management layer.

Then there is the digital library which can be a robotic system for tape or optical disks, or it can just be multiple individual devices. Moving files within the different types of storage can get complicated. Not all storage systems communicate with each other and not all applications recognize all types of storage technology. To resolve this and manage the file movement within the storage architecture there are applications called data directors. These data directors are called Hierarchical Storage Management (HSM) systems. The HSM automatically moves the content and metadata between the different storage systems. The HSM is managed by the policy and retention rules in the asset management system.

This demonstrates that there are a number of considerations and factors that need to be accounted for when designing the storage architecture. One of the first questions is: how many tiers of storage are needed?

- Online
- Near-line
- Offline
- Proxy

Online is for immediate access and high availability; it is typically used in high throughput disk arrays.

Near-line is still very accessible, but data will take more time to retrieve. This can be a digital library device like an LTO or optical disk robot. It can also be a disk array with less speed and throughput.

Offline typically means the content is on some form of removable media and sitting on a shelf tracked by the asset manager. While high-resolution media may be moved to an offline location, the proxy and metadata are always online.

Proxy is where the low-resolution version of the content that's used for browsing and creating offline EDLs is kept. In actuality, the fastest growing part of storage is proxy storage. While the proxy is low resolution, proxies live forever even after the high resolution is moved to a shelf. This includes all variations, B-Roll, and any elements created or acquired. What this translates into is that the demand for proxy storage continues to grow actually more than high resolution. Proxy or low-resolution content is not just for browsing content. Proxy resolution is the content format that feeds to social communities and mobile services. These can be short clips of news, sports, and other live events.

This all has a significant impact when it's time to consider scaling. Online or high availability and near-line storage will scale for one of two reasons: capacity or throughput. Proxy storage is all about capacity, throughput is rarely the issue.

Capacity might seem pretty straightforward; as the volume of content increases, additional storage is needed. But is this a storage capacity issue or a media management problem? Is there unnecessary high-resolution media left on high availability storage that could be moved to near-line or offline? It is important to remember that the proxy must always be available to review and access the media. Retention policies play a major role in storage usage.

Throughput is more complicated. The number of simultaneous instances and the speed that a HDD disk, SSD, or optical disk can read/write from plays a major role in how to architect the storage. Disks and solid state storage have physical limitations on how many concurrent activities (i.e. read/writes) they can perform. The number of processes any one storage node is requested to perform determines how many storage nodes are needed. While the amount of storage capacity may seem to be enough the number of storage nodes to handle the amount of concurrent processes is not. There are a number of factors involved. As there are a number of processes, the ability of the storage node to handle multiple processes is also a function of bitrate. The higher the resolution, the higher the bitrate and as a result there can be fewer concurrent activities. This pressures the need for more disks to distribute the processing load to.

In one example of this, there's a media management system that is designed for 3 edit stations, 3 ingest encoders, and 2 play-out channels. Now, there's an increase in the number of edit stations from 3 to 6 and the house format and bitrate are set to ProRes145. There may be enough storage *capacity* in the system to handle the increase in workstations, but on the performance the system slows down and cannot process that number of edit sessions at once. In this case, storage capacity may not be the issue—the throughput or how much traffic the storage can handle at once is likely the culprit. The only way to resolve this is to add more disks. In this instance the actual scaling requirement is to increase the amount of throughput and getting more storage is just a side benefit. It shows how relieving the stress on throughput impacts the entire storage architecture.

Case Study

One of the major US cable networks has one of the largest installations of a well-known manufacturer's storage. Its size is not attributed to volume, but to the number of edit stations. As the number of user edit stations has increased, their systems and processes slow down; this is a throughput problem. So to resolve this more throughput is needed and to get the additional throughput they need, they have to keep adding more disks. And each time they add a disk to manage the throughput, they get a substantial increase in storage.

Whether the scaling is a result of capacity or throughput growth it still has direct capital and operating cost implications. Disk arrays need rack space, network ports, power and HVAC, and each storage type and manufacturer scales differently. Are there more frames required, are there management nodes, what about licenses? Does the data director need more nodes and licenses?

There are a quite a few questions to ask when designing the storage architecture. Here are some of the key considerations:

- How many files are recorded at the same time?
- How many files total will be stored in each tier and for how long?
- How many different bitrates are there?
- Will the architecture be centralized or federated?
- How many types of storage are needed?
- How will the storage be segregated? (ingest, production, distribution)
- How many systems and users need to simultaneously access each tier of storage?
- What systems does the HSM need to support?
- Which network protocol is right for the design?
- What about switch ports and load balancers?
- Does the asset manager need a specific interface?

Where do I keep it? The storage system can be a designed as a central repository or in a federated network. In a central repository, there is one large storage system that is partitioned to handle all the requirements. In the central repository model all systems and applications need simultaneous access to the storage. This becomes a design consideration to the network, as well as there will be a lot of high bandwidth traffic on the network. Also, in this scenario, if the system needs to scale whether for throughput or capacity the whole system has to scale. The federated design can be in the same physical location with different storage systems servicing different processes within the entire ecosystem. In this model ingest, production, distribution, and archive are different storage systems, types, and protocols. Another type of federated design would be to have different storage systems in different locations; this allows the content to exist in diverse locations without needing to migrate large media files across the network to a central storage location. In the federated storage design, when it comes to scaling, only the part of the storage being stressed needs to scale. When designing the storage architecture, network protocols, traffic, and bandwidth management are a major consideration.

How much do I keep? Spinning disks or Solid State storage is expensive and removable media is less expensive. Retention policies are driven by business rules. The On-Demand access that users want of archival material has changed preservation policies. There are new considerations for a larger amount of content to be retained for potential use in the future, while the social communities and mobile services require an immediacy of content delivery. There is also a high demand for archival material for research, comparisons, or pure interest. For these platforms the format is typically at a lower resolution. This can mean having to retain multiple proxy sizes and formats. This also opens the question of keeping it internal or moving some of it to the cloud (this will be discussed in a later chapter).

Case Study 1

On one project the mandate of the organization is to record everything and save it forever. This translated into 22,000 hours of content per year. The question is how much of it has to be high availability vs. on demand. The archive version is the one kept forever, current content remained relevant for 30 days and could then move to near-line and once there could be moved to offline, essentially moving the removable media from the system to a shelf. In all cases the proxies and metadata still live forever in the asset management system. Any content can be searched on and restored for other uses.

Case Study 2

A different project also saves forever, however not everything. A rating system that's part of a logging system sends metadata of the selected elements to the asset management system which creates the clips that are archived and saved forever. Here there is a selective process that is trying to be practical on how much is saved vs. the amount that is actually captured. In this situation, they are monitoring the amount of content that is requested through social communities. At the same time the media is never removed from the system so the robotic digital library has expanded more than the storage.

What do I keep it on? In the preservation of media making choices and deciding on formats has been a difficult dilemma for as long as media has been archived. In the file-based world, the added dimension is not only what type of physical media will be durable and what kind of devices will be available for restoring, but also what file types and resolution will be sustainable as new file types, protocols, and containers formats are introduced. In addition to the retention policy and media type, there must be a migration policy.

- How often should the library be migrated to the next generation?
- How practical is it to keep one of each device for restoral?
- Will the next application version be backwards compatible?

It's important to keep in mind scaling and the associated costs when considering which type of technology to use for near-line and preservation storage, Near-line and preservation storage is typically on a removable media format and there are a large number of options. However if disk is a consideration, when looking at disk storage, while the cost of disks continues to decrease in proportion to their capacity, the amount of storage needed has increased exponentially. And as discussed earlier, the need to increase storage by adding disks has attached costs such as physical space, heat and power, switch ports, and mean time to failure (MTTF) considerations. How long does it take to fail? How is it protected? Digital tape is less costly, but the formats tend to sunset quickly, so there is a migration requirement and device upgrade schedule requirement. Solid state disk and solid state storage have the same questions. The physical media is fragile and device compatibility will need a migration and upgrade policy. Optical disk will have the same considerations as digital tape and solid state.

Among the key considerations in storage management is how long high resolution needs to stay online and whether a mezzanine (production resolution) version needs to be kept.

How to size storage? This is a great question and not necessarily an easy answer. Does the high-resolution content move immediately to near-line? Do users work with the mezzanine and proxy resolutions for production? How long do the mezzanine files stay online and does it have to be kept if the high resolution is being kept?

As mentioned earlier, capacity and throughput are two primary considerations in sizing storage. First, let's discuss capacity. How much new media is going to be ingested and captured on a predictable production cycle? How many different resolutions are anticipated (since file size does matter)? The table above shows the most common formats and the typical file size for 1 hour of content. To record at 4K resolution requires 3.7 TB of storage for 1 hour and 62 GB for 1 minute of video. Now look at throughput: the 4K bitrate is 3.82 GB/s, and the network bitrate is 1.03GB/s. Now add in the bitrates for the disk to read/write plus allow for buffering time in the disk/flash array. It's easy to see where a log jam might occur if the throughput of the network and storage architecture are not sized correctly. The speed of a disk is one metric of throughput, measuring how fast it can read and write. Another metric is the storage architecture. When multiple disks or Solid State Drives are mounted together, the throughput is based on both the transfer rate between storage devices and the network. This is where selecting the right transfer protocol (Fiber Channel, iSCSI) makes a big difference.

There is a lot of work being done with 4K, including applying both MPEG2 and MPEG4 compression techniques to bring down the bitrate and file sizes.

Now it's time to design the storage. Once the house file format is selected, that will be used as the reference to calculate the typical file size by multiplying the bitrate and time.

The following example calculates storage capacity first. Here are the specifications:

- MPEG2 is the main file type.
  □ One minute of DNX145 is a 1.02 GB file.
  □ At XDCAM 50Mb/s, the file is 500MB; at MPEG4 AVC100, the file is a 270MB file.
- There is an average of 4 hours of original content created or ingested per day.
  □ DNX145 is the house format of choice.
  □ The mezzanine format is XDCAM50 plus a 2MB proxy.
- One hour of PCM uncompressed 8-channel audio is 16.56 GB, making the file larger and adding to storage requirements.

One interesting note is that the way codecs are identified is by the bitrate and file size, it typically only accounts for video and politely leaves out audio and data

Format	Name	Bitrates	File Size	Platform
MPEG2 4:2:2) @ MP/HL	XDCAM	25, 35, 50Mbit/s	18 30 GB/Hr.	Production
MPEG2 4:2:2)@ MP/HL	DVCPRO	50, 100 Mbit/s	30- 60GB/Hr.	Production/Library
MPEG2 4:2:2)@ MP/HL	ProRes 422/DN×HD	147, 220 MBit/s	100 GB/Hr.	iTunes/Production/Library
MPEG2 4:2:0 @MP	DVCAM/Firewire	25 MBit/s	15GB/Hr.	Production
JPEG2000	J2K	100-250Mb/s	60-140GB/hr	
4K RAW	Red One	1.03 GB/s	3.72 TB/Hr.	Production
MPEG4	Blu-Ray	40Mb/s	6GB/Hr.	DVD
MPEG4 Part 2	H.263	700K-3Mb/s	N/A	Video Conference/Web
MPEG4 Part 2	H. 263	700K-3Mb/s	300MB	You Tube
MPEG4 Part 10	H.264	700K-3Mb/s	300MB	Web, Mobile, Flash
MPEG4	H.264/AVCHD	100Mb/s	16GB/Hr.	Production
HTML5 - Open Source	H. 264	700K-3Mb/s	Variable	Apple iPhone, iPad,
HTML5 - Open Source	Web/VP8	700K-3Mb/s		Web, Mobile

FIGURE 4-4

(VANC, HANC, Aux, CC, SAP) when representing bitrates; however, when planning storage, it is important to account for the entire file size including all overhead.

Here is the arithmetic to figure out storage for only 4 hours:

- 1 hour DNX145 Video @ 61.02GB × 4 = 244.08GB
- 8 channels PCM 24bit 96k Audio @16.56 GB × 4 = 66.24GB
  - □ **Total 4 hours DNx 145 = 310.32GB**
- 1 hour XDCAMHD50 plus audio = 42.18GB
- 1 hour proxy 2Mb/s = .99GB
  - □ **Subtotal = 353.49GB**
- Other Material @33% = 117.83GB
  - □ **Total Storage 4 Hours/Day = 471.32GB**
    - ○ **5 Day week = 3.3TB**
    - ○ **Month = 13.2TB**

Graphics, B-Roll, and effects can add up to approximately 33% of other materials, or 117.83GB in the above example. This brings the 4 hour total to 471.3GB per day; using a 5-day week program requirement increases total storage requirements to 3.3TB per week.

Using a media management retention policy that allows six months of content to be kept in online storage adds up to 79.2TB. It's also safe to assume there will be incremental growth in production materials. Plus, all proxies stay on the online storage, which adds 40% (including an allowance for overhead)—all of which adds up to 110.9TB of storage for six months of online. And this is only based on raw storage capacity. Add in digitizing legacy materials, and storage easily gets to petabytes.

To complete the planning, add in throughput considerations based on the design point for disk access times. The higher the bitrate, the more physical disks are needed. Using a disk configuration of Raid 5 with 1TB disks works out to approximately 150–175 physical disks. In addition to the disk, there are frames, power supplies, and controllers. The number of physical disks will impact other systems that attach to the storage. Each disk protocol topology differs so the type of network will determine the number of additional network ports needed.

How does it scale? What are the determining factors in scaling? More users accessing the content (traffic) will require more Input/Output(I/O) throughput, and budgeting for traffic can be accommodated by scheduling some of the file transfer processes to off-times when other processes are not working which will reduce the bandwidth and throughput requirements. Moving high-resolution files to near-line storage and managing the mezzanine files does reduce online storage; however, the proxies just continue to grow and need to remain online.

Now that the storage environment is designed and sized with adequate growth capacity and scaling considerations, it is time to discuss storage management. Storage management is based on policies and these policies control and manage the workflow within the storage architecture. There are policies that control access, availability, usage, and retention. Some of these policies are part of the asset management systems and some are within the automation system. There may be other systems that govern the way file movement within the storage architecture is managed. There are different policies for active and archived content. A migration policy for long-term archived media should also be taken into consideration.

Retention, Preservation, and Migration

There is a subtle difference between the retention and preservation policies. The retention policies manage the active content and how it moves between the different storage tiers: the lifecycle of media within the infrastructure. The retention policies define when mezzanine content is purged from the system, leaving only high resolution, when the high resolution is moved to offline archive and when the active high resolution is purged from the online and near-line to make space for new content.

The high resolution in the archive can be restored to any resolution when there is a request for retrieval. In the file-based ecosystem, the proxy is always online and can be used for search and browse, creating EDLs and requests for partial restores of archived content.

The archive or preservation policy determines the long-term life of the content. Is it really saved forever? Preservation includes the digitization of non-file or digital legacy materials. The transition from tape-based media to file includes the ability to use all the archived media in a file-based program. There are two ways to achieve this, one is to digitize all the tape-based content and the other is to digitize elements as needed or on demand. In both scenarios, the content needs to be encoded and ingested, then placed in storage, and get registered to the asset management system with metadata.

The on-demand method is more ad hoc and less costly. Using this philosophy when non-file-based content is requested that has been tracked in the library database (which hopefully has been integrated to the asset manager so that when a search is triggered, the content is identified as off-system) and in a different media form. The media manager or librarian would retrieve the content from storage (in this case from a shelf), ingest it to the system and attach the content to the original metadata tags from the library—indicating to which part of the content something new has been added. The content would then become available across the entire system with associated metadata.

A structured digitization schedule looks at the volume and types of legacy materials, prioritizing the order or value and setting up an area to ingest and tag with metadata the entire or selected library.

One of the US sports organizations implemented a file-based system and then elected to digitize all 150,000 hours of their tape-based recorded footage.

What is a migration policy? One of the practices established with film and videotape is the migration of archived media from the original media format with unknown or limited shelf life to a media format with a longer shelf life that would sustain playback capability. Another consideration in migration policy is the treatment of media located on potentially unavailable or unsupportable playback devices. An example of this is transferring video from 1", 2", and Beta tape masters to anything from D1 to D5 to DVC.

Let us now discuss file-based media. What file formats will be sustainable? What media form factors will be sustainable? In the realm of digital tape (DLT and LTO), the compression schemes and tape densities change almost yearly. The lifespan of these tapes is unknown; optical disks may have a longer life span, but the actual file format may not. Solid state and flash memory stability and longevity are still unknown.

Therefore, migration is a critical part of retention and archive policies. What is a practical cycle to adhere to when archiving to digital removable media to ensure that the content requested is retrievable, available, and restorable?

The migration policy may include a random restore sampling both before and after the migration, comparing the files and confirming their integrity. Part of the migration strategy may be to keep one of each version of software and hardware devices, enabling restoration from a variety of versions.

The digital librarian or media manager is responsible for maintaining the accessibility of the content throughout its lifecycle.

Keeping the metadata updated is essential. Migration can be an automated operation, and the asset management system may be set up to handle this. It is important that the record of the content tracks its migration if it is re-mastered or re-archived to a different format or removable media type.

Governance

Having brought up the topic of policy, it's time to discuss governance. Governance encompasses all the rules and policies that control the entire lifecycle and movement of media. It is extremely important.

These policy areas include:

- Creative
- File movement
- Integration between systems
- Rights control
- Permission and approval
- Distribution
- Media management

In all software applications and systems, there are rules and policies that control how data is managed, accessed, and distributed. This is no different in the file-based media ecosystem.

Governance establishes a set of rules and policies created by the stakeholders, who own the media and metadata. It controls the way all user groups and end users use the data. Meanwhile, each organization has its own operational structure and infrastructure. Governance enables business and production units to integrate their data, processes, and workflows to place higher value on the assets and generate new revenue opportunities.

Each process in the file-based workflow has a governing set of rules and policies. It is critical that these policies are coordinated and integrated.

These are the rules and policies that the automation processes use to move media through the media management from ingest to delivery.

Permission and approval processes manage content for distribution. Parent-child relationships are how elements within a file are tracked both separately and together.

Roles and Responsibilities

There are a number of new job definitions and operations:

- Media management
- Ingest control
- Digital library

The media manager performs a number of functions with a considerable number of new responsibilities. As the ingest manager and possibly the media logger for metadata, is this role an extension of the duplication department, that "dub guy"? Or since it is controlling the management of the media, is the digital librarian an outgrowth of the tape library? Or is this function possibly a completely new area called media management, encompassing the digital librarian, ingest, media, and metadata management AND quality assurance?

In the file and stream ecosystem of media management, there are new roles and responsibilities. From a technical perspective, this is a quality control position confirming the integrity of the files and streams as they enter and move around the infrastructure. This is where the quality and integrity of metadata and confirmation of the automated metadata entries are also confirmed.

- The ingest manager is responsible for the encoding and transcoding processes.
- The media manager monitors the movement of files between systems.
- The digital librarian manages the metadata, media storage, and access.

Depending on the size of the organization, these could be treated as a single position or, in a larger operation, media management could be its own department.

Media management is a critical piece to this new puzzle. In the IP- and file-based ecosystem, media is handled differently—with new rules and policies. Managing media requires a different tool set (i.e. asset management).

five

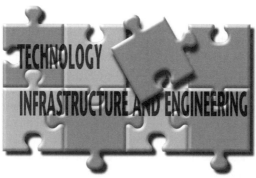

The next piece of the puzzle explores the changes to the engineering group, as well as what the engineer's considerations should be when planning and designing for IP- and file-based media. The primary role of the broadcast engineer is to plan, design, build, and maintain. The engineer has to ensure that all systems perform as they're supposed to, all problems are resolved, and everything is up and running. There are also new responsibilities for the engineer; business continuity, security, and disaster recovery.

The technology design of the IP broadcast facility infrastructure is very network, server, and storage centric. The video production switcher and audio mixer are servers located in a centralized equipment room and the controllers are only work surfaces connected over Ethernet in the control rooms. And while these devices are still proprietary technology, they are more server-centric platforms with hardware interfaces to handle the SDI/AES audio and video. All these devices have the ability to store settings and profiles in onboard storage or to removable media (USB Flash drives). Editing, graphics, animation, and play-out for distribution systems are all applications that run on servers and are connected to common storage, where the files move easily back and forth between all the different operations. Media management and metadata (for description, business, and automation) can be entered at the same time that craft production is taking place.

The core infrastructure is now substantially based on applications, servers, storage, and network. Application servers host the different tools that process and handle the media and metadata. There are processes that require proprietary or purpose-specific hardware for the application to perform. Some of these are graphics cards with their own processes and memory supporting multiple screens, encoding cards for SDI input and output and based on the type of storage topology, iSCSI, Fiber Channel, or other network interfaces. There are still broadcast products that while being server based have a proprietary hardware configuration with embedded operating systems (OS) and embedded hardware-based processes vs. software. These "closed systems" require external applications, software interfaces, and storage to integrate them into the broadcast architecture. Most of them rely on dedicated hardware- or web-based browser tools to access their monitoring and management tools without allowing access to the core applications.

In the IP- and file-based infrastructure all devices and systems are integrated and control is a layer within the Ethernet topology and integrated on the network. Where in a tape-based environment each system or device was independent and controlled using serial protocols (RS-232, 422, 485) that are point to point connections. In the IP infrastructure, the media manager and automation system are software applications that handle the transport and tracking of the file as it moves through the system. Content enters through the encoder that's controlled by the media manager and then moves into craft production where edit or graphics applications perform their roles then making it accessible to the distribution application while insuring a copy goes into archive. In the IP- and file-based ecosystem automation plays a much larger role.

Application servers move the media over the network and throughout the storage architecture. As discussed in earlier chapters there are a number of different storage configurations. In a Storage Area Network (SAN) configuration, a distributed disk architecture handles all the media coming in and going out, however, this is transparent to all users. In a Network Attached Storage (NAS) configuration, each storage location is separate and treated separately. In either of these configurations, using a separate storage philosophy instead of a central storage philosophy means there are different storage systems integrated on the network and they are organized to provide storage based on different workflows. This may include ingest storage to acquire the files, production storage to handle all of the craft, studio and remote content, archive storage, and storage for play-out.

Following the lifecycle of a file, as content is created, and a program or element is finished, it moves to the media library (storage) and becomes available for distribution and delivery to all of the different platforms.

In the IP production infrastructure the integration between systems and devices is based on networks. There are different network layers, protocols, and topologies. In the IP infrastructure, a router is a gateway and devices are connected with a switch, and the switch can be a managed or unmanaged device. This has caused great confusion when having discussion about media systems design. There is the SDI router, which technically speaking is a matrix switch with dedicated inputs and outputs. It may be managed and controlled from an application on a server or from a computer using a browser that's connected over an Ethernet network. Audio is embedded, control is a combination of RS422, and IP and intercom is VoIP.

In the IP infrastructure bandwidth management is critical. The infrastructure uses shared bandwidth in a managed network topology. Bandwidth availability and network congestion is a constant management challenge in broadcast. Media files use a lot of bandwidth, so keeping this segregated from command and control systems is crucial. Broadcast networks need multiple segments called virtual LANs or VLANs to segregate the communication between different systems and device interaction. Media, communications, control, KVM, and metadata each need their own segments. As the designer, the broadcast engineer must plan the configuration of the network.

One major consideration and concern in designing the infrastructure is to avoid or minimize latency. Interestingly, the issue of latency has more to do with transport of media, automation, communications, and command and control than it does with the syncing of audio with video. *That is not to say that visual latency (Lip Sync) is NOT a huge issue and there should be no such thing as "acceptable latency."* However, if the media (audio and video) is fully embedded and travels together, this type of latency becomes less of an issue. Back to the real concern about latency, if a control command is issued and the SDI router doesn't execute a switch or an IP automation command to trigger the playlist is late, there will be an interruption in program delivery. The same is true if the command to begin an ingest or encode is late which means the start of record is late—there will be missed content. For example, in all sports there is clock data associated with the competition and if the clock data and video are not in sync it impacts the officiating of the competition.

Latency is not the only concern, as devices communicate with each other over the network, packets from one system may corrupt the communication between other systems.

The design of the network and managing the segregation between systems is one of the key considerations. Here is an interesting description to better understand the relationship of VLANs to the broadcast system. Traditionally each of the different signal types will have different cable types and typically have their own routers (matrices). For example in a typical broadcast facility there will be routers for SDI, AES, MADI, Intercom, and RS422/232 plus an Ethernet switch.

Looking at the infrastructure bandwidth, the next generation of SDI is 1080P and 3Gb/s, which will also enable it to handle new formats with higher bitrates, like 2K, 4K, and 8K. Manufacturers of audio video routers and terminal equipment are pressed to provide products that will support 3G. In the IP world, the next generation is 10Gb/s, 40Gb/s, and 100Gb/s which in comparison is much bigger than 3Gb/s.

In the IP architecture, each of these signals either is encoded to IP or originates as IP. Therefore they all use the same cable type (CatxE or fiber) and there is a common network switch. In the IP architecture they are differentiated and segregated by using VLANs. Multiple VLANs can co-exist in a single network, on the same cable/fiber and on the same switch. The switch is managed and controlled so that each signal is optimized and maintains the integrity to perform its function correctly. There are different ways to manage network traffic and the segregation of VLANs, Access Control Lists (ACLs), and Trunking will be discussed a little later in this chapter.

Signal	Cable	Connector
SDI Video	Coax/fiber	BNC/ST
AES/Madi Audio	Coax, multi-pair/fiber	BNC/Phoenix/ST
Analog Audio	Single/multi-pair	XLR/Bare end
Machine Control RS422/232	Multi Conductor	DB9, Phoenix
Management	Cat5/6e	RJ45, Punch block
Intercom	Single/multi-pair	XLR, DIN
Router Control	Coax/multi-pair/CAT5/6e	BNC/phoenix/RJ45, DB9

FIGURE 5-1 Non IP signals, cable and connector types

Signal	Format
Video	MPEGTS, JPEG2000, H.264
Audio	Madi, CobraNet
Intercom	VoIP
Command and Control	xml
Machine control	xml
Router control	xml

FIGURE 5-2

So in the transition to an IP-based core infrastructure, as media is encoded to IP, the output of these devices is now a network port that goes to a port on a network switch. Each port on the switch can be configured for the signal it is transporting and isolated if necessary. There is now only one cable type, two actually since fiber is used for higher bandwidths and for sending the IP over extended distances.

So in essence, each VLAN can be compared to the unique cables previously used for each of the signal types. Now, each of the signals are in the same switch and groomed into a single IP stream that travels over the same cables/fibers. As it connects to each device, the device understands which VLAN and signal it is taking commands from, sending or receiving files and what process to execute or file to manage.

Planning and designing the IP infrastructure requires knowledge in network architecture, an understanding how to configure applications, servers, and storage. Most of these devices are managed and operated with keyboards and mice- or touch-controlled Graphical User Interfaces (GUI); they no longer have physical controls on the front of the devices or use specialized remote controls based on serial (EIA RS-232, 485, and 422) and other proprietary protocols.

In the design of the IP broadcast center, the server hardware is centralized, and extenders are used for Keyboard, Video and Mouse (KVM), this could easily require multiple keyboards, mice, and monitors at every workstation. To avoid this congestion and keep cabling at a minimum in control spaces, there are KVM switches and matrices. This allows multiple users in multiple locations to access the applications and servers remotely and from a single workstation. This is similar to having multiple SDI router control panels or using a RS-422 machine control router to facilitate more than one user or control room accessing a pool of sources. The KVM extender uses an IP transport and travels over the network topology as a layer. Actually because the KVM provides the main user display, this layer requires high bandwidth for the high resolution video feeding all the displays. This puts a tremendous load on the network and the KVM does not require interaction with any of the other LAN segments, therefore keeping it isolated in its own VLAN is critical or it will cause issues (latency or collisions) with other network traffic. One Keyboard, Mouse and Video display screen can control hundreds of servers and network devices; in addition multiple KVM stations can access multiple devices from multiple locations; the user can toggle between servers. While many IP-based systems are managed using a web or browser interface, there needs to be a computer to open the browser, in the case where the devices have dedicated applications that run on servers or workstations with software interfaces called dashboards, the KVM is used to access them.

The following case study is a good example of how using KVM provides power and flexibility in a production environment.

Case Study

A large global organization has 25 meeting rooms plus 5 production studios, each with HD cameras and multiple channels of audio that are recorded to file and broadcast live. These rooms and studios all operate at the same time. There are over 3000 IP addressable devices with 300 of them using the KVM matrix for management and control. There are over 50 locations for operator control positions, each position having a single KVM station in addition to the other control surfaces to execute production. The technicians use the KVM to toggle between systems and applications in order to record and manage the content workflow from each meeting room and studio.

Changes in Engineering and Maintenance

The interaction between business units and the handling of media has changed many of the workflows and business processes. Business models have also significantly changed for creating and distributing media. As media moves throughout the entire infrastructure, it needs to be accessible to all departments/ units including production, distribution, library, legal, finance, marketing, and business intelligence. These groups need access to the media to input and export metadata and to ensure that the appropriate business rules, policies, and permissions are in place so they are protected as the metadata is distributed into the marketplace.

In the file-based and IP architecture, that premise remains the same; however, the engineering department now needs additional knowledge and skill sets. Previous puzzle pieces described changes in production workflows. This piece covers the challenges and opportunities the broadcast engineer faces when designing for file-based and IP workflows particularly as an evolution from a tape-based SD/ HD-SDI facility.

Traditionally, IT engineers and managers have had different priorities for broadcast/production engineers. The requirements of an enterprise IT architecture are substantially different from those of the broadcast IT architecture. Both at their core build on servers, applications, network, and storage. Both use commodity hardware and software including operating systems and databases in addition to proprietary hardware and software. However, there are significant differences in the types of applications and system configurations, and the two have very different priorities.

It's time to highlight some of the differences between **Enterprise** IT and **Broadcast** IT engineering.

Broadcast is all about getting on the air and staying on the air. **Enterprise** is about security, communications, accessibility, web services, and storing documents. Enterprise has the luxury of taking systems offline for routine maintenance. Broadcast does not have that luxury.

The first and foremost difference is system availability. The broadcast IP infrastructure has to be online and available 24/7 with no downtime. Bandwidth management in the broadcast IP network is critical. On the enterprise side, bandwidth management is less critical, except when users are watching movies online and interfering with business operations.

One of the most common differences between broadcast and enterprise is the need for urgency in solving problems. In the enterprise IT world, problems are solved by calling the help desk, opening a trouble ticket, and waiting for a response. Other than catastrophic failure, there is typically not the same sense of urgency in addressing enterprise problems. And when it comes to diagnosing and solving a problem or disruption in service, the Enterprise IT approach is to first forensically understand where the problem came from before restoring service. In the broadcast environment, if there is a disruption in service, the highest priority is to identify the problem, restore the service, and ONLY then, fully diagnose and solve the problem so it doesn't interfere with services again.

In the broadcast world, the engineer knows that systems need to be working and staying on the "air" is the most important responsibility. In broadcast there is NO DOWNTIME!

On the enterprise side, routine maintenance schedules take systems off-line. In broadcast, that is not really an option; services cannot have interruptions, so maintenance needs a process to work around production and distribution schedules.

The traditional role of the broadcast engineer is to design, build, and maintain all systems and equipment. Simply stated, their responsibilities are to keep everything on the air and ensure there is no interruption of broadcast and production operations.

The broadcast engineer is now also responsible for multi-platform integration and systems administration in the IP ecosystem.

In each of these areas, the IP- and file-based architecture is substantially different from the infrastructure of an SDI facility. In the migration from analog to digital to SD-SDI and then HD-SDI, the core principles in the architecture stayed the same. The core of SDI systems is the audio/video router with the production devices being connected using the usual combination of distribution amplifiers (DA), Analog to Digital and Digital to Analog converters (A/D and D/A) plus Frame Synchronizers, cross converters, and the rest of the "technical glue" technologies. It's interesting that the management of these systems is through a proprietary software application or a browser interface accessed over the network.

In the IP ecosystem, the core of the broadcast infrastructure is *Ethernet routers and switches*. IP is a duplex path with audio, video, command, and control; communications are on the same stream, cable and port on a device. The devices are encoders and transcoders, servers and hard drives. Incoming signals are IP or encoded to IP or as a file transfer.

Quality control has changed; SDI signals are monitored using digital waveform monitors and VectorScope. Quality control now includes Quality of Service (QoS) and Quality of Experience (QoE). The engineer needs to understand the way traffic moves across the network, as well as the requirements for each delivery platform.

Planning for the Future

When planning and designing a facility, it is critical to try and anticipate growth. The core technology infrastructure is different and has several new considerations:

- Extensibility
- Scalability
- Sustainability

Extensibility: A design principle where the implementation takes into consideration planning for future growth.

- Are the systems being designed anticipating growth?
- Is there enough spare capacity in the base design?

With technology changing as rapidly as it does, it is essential in the planning and design of a facility to assure that extensibility provides for and *anticipates* future growth.

Scalability: This is the principle of selecting technologies and designing the systems that allow for incremental growth without needing to replace entire systems.

- Is the system able to grow and how?
- What is the growth based on?
  - Is there an increase in the number of people or systems that need access?
  - Is it servers, storage, network, or application licenses?
  - Are there enough ports on the switch?
  - What about bandwidth on the network and in storage?
  - Are there more content production, distribution channels, or delivery platforms?

Scalability is viewed slightly differently in the IP architecture. The core design of a traditional broadcast and production facility is centered on the capacity of audio and video equipment, terminal and distribution equipment, and size of the A/V router. AND most, if not all, engineers will agree that routers are always too small (not enough inputs) by too small (not enough outputs). In the SDI (baseband) environment, the task of adding an edit suite or an additional VTR for production or transmission was never a small task. There are many connections to infrastructure and terminal equipment with different cable and connector types that are all hard wired to specific devices.

In the IP architecture, scaling, growing, or expanding is accomplished by adding servers, storage, and applications. Are there enough network ports? Is there enough bandwidth? What about throughput?

Sustainability: This has as much to do with Total Cost of Ownership as it does with maintenance.

- Keeping technology productive and available over its usable life
- What is the lifecycle? Does it have a sunset timetable?
- How is it maintained? Hardware vs. Software?
- Is there a service level agreement (SLA) with the manufacturer?
- When is the decision to replace vs. repair/upgrade made, with so much hardware and software changing and going out of date so quickly?

Maintenance is a different beast in the IP environment. Maintenance was always caring for a broad spectrum of equipment (e.g. cameras, switchers, monitors and mixers, routers, and terminal equipment) and keeping a cadre of spare parts and consumables on hand. The engineer now faces additional responsibilities with the maintenance of software, servers, networks, storage, and digital libraries. There are new little black boxes for adapting devices, converting signals, and interfacing systems.

Even spares have a different concept, there are hot spares and cold spares. Cold spares sit on a shelf or can be integrated and wired in a rack and powered down. In the event of failure or maintenance the cold spare is brought online and the primary taken offline. Hot spares are devices that are online and configured to failover in the event of a problem. This could be an extra card in a frame or a redundant server. There is also hot swappable, which is the ability to remove a failed component and replace it with a working component without power cycling the system. Thinking about what types of spares need to be kept on hand is an open and constantly changing question. How systems are maintained has changed substantially. This includes:

- Software patches
- Software upgrades
- Bug fixes
- Hardware
- Cold spares vs. hot spares

Total Cost of Ownership

All businesses maintain budgets, and engineers are typically required to create capital and operating budgets to support a facility. Total cost of ownership (TCO) is all the costs associated with operating a device or system.

- Annualized maintenance costs for software and hardware
- Hardware replacement vs. upgrade
- Software version upgrades
- Sunsetting hardware and software
- Infrastructure – Electric, Environment

When we look at total cost of ownership, some of the costs formerly associated with spare parts inventory are now allocated to annual maintenance called service level agreements (SLA) for hardware and software. This can be based on as much as 15–18% of the original cost of the product. When looking at TCO there is a higher probability that the hardware will be replaced long before it actually expires.

When a vendor releases a new software version or a new application, they may require different hardware, faster processors, more storage, or more memory. This means that the software application and the hardware it runs on, which was perfectly fine yesterday, now needs an upgrade! And that server that is only two years old and running fine is not perfectly fine for the new application. We are now seeing that, long before a device reaches the end of its lifecycle, there is the likely scenario of a potential upgrade or replacement. The product lifecycle of servers, disk, and flash and digital tape storage is different from the lifespan of a video tape machine and magnetic tape.

Another major difference in the transition to software-based systems is that vendors are upgrading versions very quickly, and they are releasing completely new versions that are more substantial than a new feature update or a bug fix. As vendors and service providers introduce new versions with new features and new functionality, they very strongly encourage users to buy them as they abandon support of the previous versions. This is gracefully described as "sunsetting."

This creates a serious dilemma. There may be no requirement for new features, and while there is nothing wrong with the current software version or the hardware that it's running on, the vendor may no longer support it, despite the maintenance fees they are charging for service. Hardware and software are reaching their end of product life on much shorter cycles than traditional broadcast equipment ever did.

For the engineer, building the IP infrastructure means installing and configuring software onto servers, using remote access to configure switches, and being familiar with storage architecture to set up partitions based on user requirements and manufacturer specifications. This is in addition to the specialty hardware and software for media services, like devices or adapters for encoding, decoding, transcoding, splicing, and grooming.

The broadcast infrastructure is still very much a hybrid of SDI and IP. The major studios and remote production units that produce programming continue to deliver it on HD digital tape, as do advertisers, so the broadcast engineer has to maintain a way to manage the tape and its content as it is being encoded to files. The design needs to include interfacing all the software applications.

The IP architecture still needs maintenance; quality control now includes packet loss analysis and bit error detection. Software maintenance mostly includes updating service contracts, installing patches and upgrades, or finding a bug that is causing service disruption.

Hardware is a bit different. Hard drives in servers can be replaced; any devices that have cards or blades can be replaced. Power supplies still fail. Typically in server, storage, and switching products, a repair is considered swapping out a card or disk. Thus, the broadcast engineer is also a network and software engineer.

The design of the IP architecture includes delivery to multiple platforms. This encompasses formats, firewalls, APIs, and software interfaces.

The broadcast network has a controlled interface to the enterprise network. The broadcast engineer is the systems administrator for the broadcast network and needs to interface with the enterprise system administrator to maintain seamless transactions across the two networks.

The IP infrastructure is a multi-layered topology of networks:

- Media Transport
- Automation
- Command and Control
- Management
- Communication

The core network is based on Ethernet. The storage uses a variety of different network protocols and interfaces (e.g. Ethernet, Fiber Channel, iSCSI, and Ultra-Wide SCSI). These are closed network layers that move large media files between servers and have storage at higher bandwidth without adding congestion to the core network. All signals ultimately travel over the same IP topology. There are many layers within these layers; the Ethernet layer has the VLANs that manage audio, video, communications, command and control, automation, and management. Each layer in the network has different performance requirements, and it is the engineer's responsibility to design and build the network in a way that supports this.

The design should consider the relationship between the network segments and how the content that travels to and from the core business and production networks serving Over the Air and Internet delivery. Each of these network segments must be protected and isolated. This can be done by using firewalls between each of the business units, managing the crossover between the enterprise IT and broadcast network.

There is overlap in many areas; at times, the two engineering groups need to work together to provide continuity in operations while still maintaining the necessary separation.

The firewalls control access and regulate bandwidth. It is interesting that, while it is important to keep the heavy media traffic on the broadcast network, there is a consistent growth in streaming media access as well as services like video conferencing and IPTV that are on the enterprise side of the network. These growing services are creating new issues of network congestion and bandwidth usage.

The two IT groups have different knowledge and skill sets that need to come together to design the IP topology for a media-centric production environment. The broadcast engineer can identify the critical areas where latency and QoS cannot be compromised, and the enterprise IT engineer can identify where traffic on the network for business continuity and security cannot be compromised. A robust network plan can be easily created with all the requirements and parameters identified in the planning stages. The enterprise network engineer has experience in VLANs, trunking, and creating the access control lists (ACLs) that will be critical when configuring the broadcast network.

The enterprise IT group typically has the primary Internet connection that controls all remote access and provides any wireless services. The IP broadcast center now more closely resembles a data center. This brings with it the need for engineers to develop new skills. The new tools of the trade are a keyboard, mouse, and monitoring systems that are browser-based dashboards. Meanwhile, the days of the greenie and tweaker are sunsetting like so many other things.

Broadcast and production technologies are based on applications, servers, storage, and networks. The processes and technologies that are acquiring, transporting, managing, transforming, and delivering media are all applications that run on servers. The engineer has to deal with operating system (OS) updates, and application upgrades, and being concerned about compatibility. When an application gets updated or upgraded, they must know how many other applications are affected and whether there is an impact on hardware and/or device interfaces.

Media and metadata are organized in digital libraries that live in the storage architecture. These are hard disk arrays with multiple servers managing the disks. There is middleware (like a data director) that handles the communication between the media applications and the storage architecture. All these technologies are integrated over different kinds of networks. The digital library can be a robotic tape device.

A network analyzer monitors this infrastructure; there are management tools for each application and service.

Once the design and build are complete, the engineer is responsible for keeping the facility and systems functioning correctly. This includes testing and monitoring to maintain signal quality.

There are a number of test and measurement tools that are used. When the transition to HD/SD-SDI first occurred, new monitoring tools were introduced and traditional waveforms changed to digital waveforms. And with all the embedded information, the monitoring needed to be able to confirm the integrity of the entire signal.

In the production environment, there is still a video operator that needs test and measurement tools to shade cameras and assure quality. The audio engineer has broader responsibilities with multi-channel programming and uses software-based tools for test and measurement.

In master control, systems are typically monitored by alerts and alarms that make a sound or light up. In the IP-based ecosystem, alerts from software-based systems have become colored icons on screens and can send an email or text message. The engineer needs to make decisions on what type of alerts and alarms to design into the IP system, as well as how the response is managed. Many systems can be accessed remotely by a browser or KVM. The engineering shop looks very different these days: it is now computers, applications, and multi-view displays showing dashboards of servers. File and stream analyzers are applications on servers.

Transmission is just one of the many distribution chains and is much more than an RF signal. Transmission monitoring extends beyond spectrum analysis. Over the air transmitters are now fed by an ASI or IP stream that carry multiple channels embedded in a single stream. Files are delivered to On-Demand broadband, cable, satellite, IPTV, and mobile service providers. Streams are delivered to content distribution networks (CDN) for broadband and mobile distribution.

In broadcast, test signals and monitoring tools were created to ensure that all systems and devices could be calibrated to the same set of parameters and meet their performance specifications. In the digital and IP environment, *nothing has changed philosophically*; test signals and measurement equipment have simply evolved to address SDI and IP.

Interestingly enough, the broadcast industry still uses a variant of color bars as the basic setup for video color reference and levels, as well as a "pilot" tone for normalizing audio. There are new video test signals and patterns used for testing MPEG and SD/ HD-SDI with better audio test signals for normalizing systems. In IP and file broadcast and production systems, there are NEW devices and systems to monitor with new ways to test and measure them. Media needs to be tested prior to and post encoding and compression. In the IP- and file-based ecosystem, testing files is different than testing streams.

SD/HD-SDI and surround sound brought a new layer to engineering skills and knowledge base. Test and measurement for SDI introduced Eye Waveforms, Gamut, Jitter, and Timing as waveforms for analyzing SD/HD-SDI signals. While SDI is considered uncompressed full bandwidth video, it is an MPEG2 stream with multiple channels, there are multiple AES audio channels embedded in the stream and the vertical interval is nonexistent. However there are still many of the same services that need transport within the main signal stream. These services and data sets that used to go into the vertical interval, on Line 21 and other places in the NTSC signal, now go into ancillary channels within the MPEG profile. These are called the Horizontal Ancillary (HANC) or Vertical Ancillary (VANC) and Auxiliary (AUX) channels within the MPEG stream. SDI is managed very similarly to the way analog was with discreet monitoring and measurement devices. There are multi-viewers that show many program streams in a multi-screen display. However the test and measurement tools still work on a single program stream, the output of the test equipment can be shown on the multi-viewer.

There are new waveforms and the engineers and operators use these waveforms to monitor and manage the audio and video levels.

There is component gamut, vector and eye waveform, plus the monitoring tools include analysis data showing cable loss details as well as other video analytics (e.g. Jitter, Gamut, and MPEG resolution). The engineer needs to read these waveforms, understand what kind of defects or anomalies would show in the MPEG SDI stream, and then determine what steps and devices could correct them.

Moving to IP- and file-based systems introduces a new set of skills and knowledge base for the broadcast engineer. IP streams over Ethernet and other transport and is monitored differently. The characteristics of IP- and file-based video are different than SDI. For IP streams and files, SDI tools do not represent what is actually going on. True, they show a decoded representation of the signal, but they do not show the IP stream or file in its native form without introducing their own artifacts.

In the IP media architecture, the network has an equal if not greater role than the routing and distribution architecture in SDI. IP media moves as streams, or files in packets, between devices over a network. The new tools in the engineer's kit include packet analyzers, bit error detectors, network bandwidth analyzers, and packet loss detectors.

Meet the new broadcast engineer: his tool chest is a laptop, and his monitor is a KVM device with access to a variety of software applications that measure and validate the IP streams and files, devices, and network.

Some of the characteristics the stream analyzer looks at are encoding errors, bit error, and packet loss. For file analysis, there are other parameters that are analyzed (e.g. checksum, syntax errors, frame rate, freeze frame detection, audio loss, and audio loudness).

In addition to monitoring stream and file quality, the broadcast engineer needs to monitor network traffic; the Studio to Transmitter Link (STL) is now an Ethernet circuit. The encoding and transcoding processes must be monitored to prevent artifacts or latency from being introduced to the stream or file. These analysis tools for IP and files are applications running on servers on the network. One consideration in configuring the analysis tools is the impact they have on network performance and access to files.

File analysis tools can run faster than real-time. These tools may add overhead to the ingest time, whether it's an encode or file transfer. Stream tools run in real-time. Both are a heavy burden on the processor, so it will need a dedicated server.

Broadcast engineers have different priorities from those of IT engineers. Both are responsible for ensuring that the core infrastructure supports business operations. The broadcast engineer's priorities are quality assurance of the content, data, and delivery—getting on air and staying on air. There is NO acceptable downtime. The enterprise engineer focuses on network traffic, systems backups, and security. In contrast, taking systems offline or out of service for routine maintenance is typical for the enterprise IT engineer.

Latency is a major concern and should be an important consideration. While the broadcast architecture has become IT centric, it still requires a media-centric engineering philosophy and mind set. IT engineers and managers have had different priorities from broadcast/production engineers.

Traditionally, network latency, synchronization, and timing were of no concern to IT, as they do not impact business operations or the performance of business applications. Broadcast engineers understand the needs of media, while IT engineers understand servers, storage, and networks. Both are valid, and it is important to recognize the differences and how to support each other.

Both broadcast and IT engineers are concerned about stability, integrity, and uptime.

Broadcast operations are 24/7/365. This presents major challenges for routine maintenance and taking systems offline. Network congestion is not typical when moving documents or spreadsheets and accessing databases. Now, enterprise IT engineers are challenged with streaming media, online webinars, video conferences, IPTV, and Skype. Latency comes in many colors and flavors, such as files not arriving, command requests being delayed, as well as the most well-known latency between audio and video, introduced in the encoding or decoding process referred to as LIP SYNC.

More than in enterprise IT, in broadcast IT there is considerable automation to move files around. On the enterprise side, a document may be shared or multiple users may have access to a shared database within an application. In media, the file actually moves locations between ingest, archive, production, and distribution. There is automation in all aspects of the IP architecture and workflow. There are different types of automation that perform multiple operations. Automation includes software applications that interface with other applications running on servers or dedicated devices.

On the enterprise side, a document doesn't change format. A Word document stays a Word document or may become a PDF. On the broadcast side, not only could the format change, but also there are myriad versions based on distribution platform. For example, a .mov or .qt may become an MPEGTS, MXF, GXF, or LXF and then an .flv, Mp4 or .wmv—and at different bitrates from 250Mb/s to 500Kb/s.

Broadcast engineering in the IP and file ecosystem opens new opportunity for skills and knowledge. In addition to the full complement of HD-SDI systems and devices, there are new devices and new architecture. The broadcast engineer is re-defined. They are now responsible for:

- Applications and OS installations
- Software configurations and management
- Database management
- Software and hardware maintenance and upgrades
- Network engineering and management

Media is encoded to a stream or file either directly out of the camera or in a production control room. From there, it enters the IP and file ecosystem, where engineering and quality control change. In addition to audio and video production devices, engineering responsibilities now include servers, storage, network, applications, and database management.

It is fair to say that all broadcast and production equipment is essentially computers and/or servers. Even the control surfaces and heads of SDI and AES production devices are IP-based processors. Cameras have onboard processors and even lighting instruments have smart controllers built in. Microphones are one of the few holdouts.

As in all broadcast operations, these devices need to run 24/7 and be maintained without interfering with operations. The broadcast engineer is now also a network administrator, responsible for VLANs, SANs, NAS, and WANs. Quality control analysis tools are applications running on the network. In the IP and file ecosystem QoS (Quality of Service) and QoE (Quality of Experience) are the new measures of performance acceptance.

Now let's look at the integration of business systems into the broadcast and production systems. Figure 5-3 shows, from the engineering perspective, which systems are under automation, where they interface to other systems, and how they integrate to form a whole architecture.

Other pieces to this puzzle captured the crossover between enterprise IT and broadcast, where the systems are intermingled and need to be able to communicate seamlessly. As the picture of this new puzzle takes shape, these pieces look at the workflow integration and at the way the business (enterprise) network and

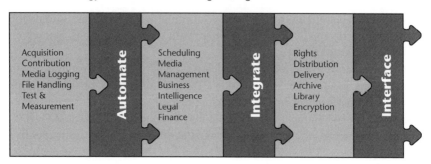

FIGURE 5-3

broadcast network integrate within the infrastructure. There is no clean separation between the broadcast and network segments; there are user groups and subsystems that need access to both broadcast and enterprise.

It is important to take a big picture perspective of the systems and subsystems when designing the IP infrastructure. It is imperative in building a new maintenance philosophy to understand all the hardware and software interfaces in the various production and enterprise systems.

The changes in the engineer's roles and responsibilities for designing, building, and maintaining the broadcast facility is another piece of this complicated puzzle.

six

TRANSMISSION AND DELIVERY

The next set of puzzle pieces are transmission and delivery. Previously, whether sent to an over the air transmitter, cable, or satellite provider, the content was the same. There were differences in the programming but not in the format or media type. Not anymore.

Today, there are different delivery platforms and various distribution methods. Each has different considerations to meet the new delivery requirements. There are a number of different paths for distribution and media is formatted differently for each platform. Platforms today include Over the Air (OTA), Over-the-Top (OTT), web stream, web On-Demand, cable and satellite with On-Demand, smart TVs, tablets, smartphones, and gaming consoles.

Transmission includes IP as the new STL and the interconnection between different locations across the network (e.g. affiliates, station groups, production centers, and remote venues like sports arenas and stadiums).

The entire media industry is focused on the "second screen" and how to capitalize on it. There are some that consider non-traditional TV devices the first screen—an interesting concept that will be explored a little further into this chapter.

When media is delivered to smart devices, the consumer is expecting additional information and services. Metadata is not just for managing media within the production workflow.

So, is TV—or better said, the large screen—a thing of the past? What about cutting the cable?

Actually, it's not likely. It is certainly true that the term "TV" or "television" (meaning a one-directional channelized display device) has changed and now we use it more to describe the type of program content we watch.

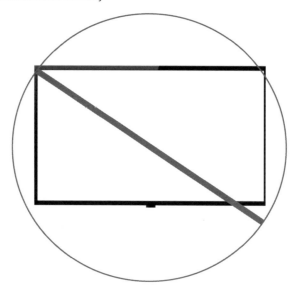

FIGURE 6-1

Before we discuss multi-platform delivery though, let's look at a few important terms:

- **Television (TV)** is a telecommunication medium for transmitting and receiving moving images with or without accompanying sound.
- A **television set** (also called TV) is a device that combines a tuner, display, and speakers for the purpose of viewing television.
- A **television program** (also called a television show) is content that is intended for broadcast on television.
- A **broadcast network** is an organization that provides live or recorded content, such as movies, newscasts, sports, public affairs programming and other television programs for broadcast over a group of stations.
- **Broadcasting** is the distribution of audio and video content to a dispersed audience via any audio-visual medium.

Looking at the above definitions, it's reasonable to say that streaming media and webcasting can now be accepted as a form of broadcasting despite the lack of physical broadcast stations—in which case, its service providers COULD potentially be considered broadcasters or even broadcast networks.

This may be a bit of sacrilege! Cable and satellite networks have been accepted, so the leap of faith to acknowledge that broadband, online, On-Demand, mobile, and portable media delivery systems are also networks shouldn't be that complicated.

The broadcaster or the program origination center delivers content to all these different types of networks, each having its own set of parameters and production requirements. In addition to the main program content, electronic program guides (EPGs), descriptive metadata, rights management, and protection are all companions to the programming.

The second screen is an addition to the viewing experience that is being used to augment and enhance user experience. Metadata is a valuable tool necessary to monetize delivery to the second screen. Using sports as an example, fans are interested in getting statistics and player information, possibly playing a fantasy version of their favorite sport in a social community while watching the live sporting event. Being able to access all this is based on what metadata is included with the programming.

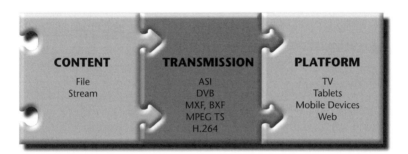

FIGURE 6-2

The basic concept of transmission and delivery is the same in the IP and file architecture. It's only the technologies that are different. Here is a good place to look at cutting the cable. This is the concept of using Over-the-Top and Internet services to view content, bypassing cable, satellite, and IPTV service providers. It doesn't quite work that way. While online services are moving into the content origination business (i.e. YouTube, Amazon, Netflix, etc.), one of the first questions to ask is—how do users get their Internet? Hmm, cable, IPTV, or telephone company. So similar to when cable and satellite were introduced and it rang the death knell of the broadcaster, that didn't quite happen. If users start watching programming from different program originators using Internet, the same service providers, i.e. cable and telephone companies, will change their rate structure and begin billing based on bandwidth usage. So it's not really cutting the cable, more like changing it.

Back to transmission and delivery, once content is ready to leave the broadcast center, it is formatted for each of the different platforms. It will leave the broadcast center as a file or stream as ASI, IP, or as files in multiple formats, with protocols such as RTMP, RTSP, HTTP, and MPEGTS.

Many of these platforms use the same protocol:

- ASI and IP transport streams feed digital transmitters, cable, satellite, broadband, and IPTV.
- RTMP Adaptive and Smooth Streaming and HTTPLive Streaming all feed to the web, tablet, and smartphone.

Files, however, are handled differently. Typically they are delivered to a content distribution network (CDN) that handles all the end user requests, authentication, replication, and load balancing.

Transmission is also about contribution and remote production. There are technologies that can multiplex cameras, that transport the ISO feed, provide return video and communications with the production control room back at the broadcast center, where it can be fully controlled, integrated with all the broadcast systems, and produce a finished program for broadcast or live. In a field production, where the media is captured to a disk, a server, or even a laptop, if there is sufficient bandwidth or a dedicated return path (backhaul), all the media is uploaded back to the broadcast center. News crews will use the open Internet and with technologies that can bond broadband that enables them to get enough bandwidth to transmit over cellular networks.

This raises other questions:

- Where is the media going and how does it get there? As a file or stream?
- Which container or wrapper will be used—MXF, LXF, BXF, QT?
- What about file size vs. bitrate? How many types of formats?
- Does it need protection? If so, what kind—DRM, watermark, or encryption?

It is also important to look at where the media is going.

- What type of client device and what type of delivery bandwidth are available?
- Is it contribution to a station or affiliate?
- Is it going Over-the-Air, Over-the-Top, or through broadband?
- Is it a file or stream?
- Is it compressed or transcoded to a specific file size and/or bitrate requirement?
- How many different bitrates are needed? (This is based on browser and player types PLUS client bandwidth limitations for distribution.)
- Is it watermarked for tracking and clearances?
- What about metadata elements and encryption for managing and controlling subscriber access? (This is next generation conditional access.)

These are all decisions that need to be made and established as a set of rules that govern distribution.

The format decision is based on the workflow and internal needs of the organization. There will be a consistent change in delivery formats. H.264 ~ H.265 and more on the horizon . . . MPEG7 and MPEG21 are in committee.

While delivery formats are striving for smaller files and streams, production is looking towards 1080P, 4K, and 8K as production resolutions.

Resolution standards should be based on the entire workflow; one resolution does not address all needs.

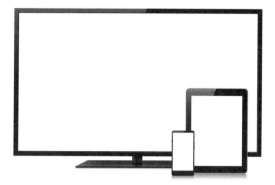

FIGURE 6-3

The content delivery landscape is fluid. We are constantly expanding the number of platforms and devices that are getting content and making content useful and device appropriate so that it can be monetized.

- Over-the-Air (broadcast) and mobile DTV
- Cable, satellite
- IPTV
- TV everywhere/Over-the-top
- Mobile
- Broadband Live and On-Demand
- Portable players—tablets, wearables
- Media consoles—Roku, Apple TV, TiVO, VuDu, etc.
- Gaming consoles—Wii, Xbox, PS3, etc.

It is important to mention that reaching the maximum number of end users is the Holy Grail of all program creators and that's what leads to the second screen. Each of these platforms has its own specification for playing media in addition to a different production requirement to maximize the effectiveness of the platform. There is some commonality with variations for each platform when it comes to compression and delivery formats. When Over-the-Air became digital, it enabled multiple channels in the same spectrum. Now the mobile DTV standard which enables mobile devices to receive Over-the-Air has been approved and stations are upgrading their transmitters to enable the addition of mobile channels to mobile devices with ATSC tuners. Cable, satellite, and IPTV are using set top boxes (STBs) with DVR capability. This uses the STB with OTT as the player instead of streaming and using bandwidth. At the same time, service providers are offering bundled services with wireless services for second screen delivery within the same premises as the STB. Portable players can download media, while gaming consoles, phones, tablets, and web can both download and stream.

Case Study

In addition to producing content in the broadcast center for multiple platforms, US sports producers are using live cameras at events and streaming directly to online distribution. In addition, technology providers are building integrated production in box systems that can pull from wireless and mobile devices into a production switcher/mixer and create finished production switching live between different wireless devices and send stream directly to a CDN.

Format	Name	Bitrates	File Size	Platform
MPEG2 4:2:2) @ MP/HL	XDCAM	25, 35, 50Mbit/s	18 30 GB/Hr.	Production
MPEG2 4:2:2)@ MP/HL	DVCPRO	50, 100 Mbit/s	30- 60GB/Hr.	Production/Library
MPEG2 4:2:2)@ MP/HL	ProRes 422/ DNxHD	147, 220 MBit/s	100GB/Hr.	iTunes/Production/ Library
MPEG2 4:2:0 @MP	DVCAM/ Firewire	25 MBit/s	15GB/Hr.	Production
MPEG4	Blu-Ray	40Mb/s	6GB/Hr.	DVD
MPEG4 Part 2	H.263	700K-3Mb/s	N/A	Video Conference/Web
MPEG4 Part 2	H. 263	700K-3Mb/s	300MB	You Tube
MPEG4 Part 10	H.264	700K-3Mb/s	300MB	Web, Mobile, Flash
MPEG4	H.264/AVCHD	100	16GB/Hr.	Production
HTML5	Open Source			Apple iPhone, iPad,

FIGURE 6-4

The decisions on codec, bitrate, and container impact many systems and processes in the IP architecture. Some of these decisions are driven by quality, technology choices, and budget, while others are driven by the final destination of the delivery channel. CDNs have specific requirements for contribution before they will process for accessibility.

Files are handled differently than streams in distribution. Most delivery platforms use embedded media players like Flash, HTML5, HTTP-live, and Silverlight. For web delivery, most browsers will handle all of these. Services like Netflix base their streaming service on Silverlight, while Amazon uses a proprietary format. There is considerable disruption with content providers. Flash had been the most common format, but that's changing. For iTunes, Apple requires an 88-220Mb/s ProRes file, while others use ON2, MXF, or QT (Mov). While this book is advocating for establishing a single format within the production center for managing media, distribution will still require transcoding to address the formats for different platforms.

There are multiple transport protocols used, such as ASI, IP (TCP and UDP) over Fast Ethernet or MPLS (typically over fiber that connects the broadcast center to the transmitter and affiliate stations). DVB-S ASI was first used for satellite delivery, and now with DVB S2-ASI, DVB S2-IP, and DVB-T, they are used for both contribution and distribution over terrestrial circuits as well.

There can be multiple formats that travel over these transport protocols, such as JPEG 2000 or J2K, MPEG2 TS, and MPEG4 H.264, and H.265, to name a few. The number of different platforms complicates program origination. Commercial integration is more complex as well. Third-party services handle the user-specific ads and promotions that the digital platforms enable.

The broadcast center receives and originates content as high-resolution files and streams. This is typically over dedicated fiber or satellite. The master control network origination systems and operations transcode from the high-resolution format to the myriad delivery formats and protocols. This is where an automation system is important for managing the transcoding and distribution.

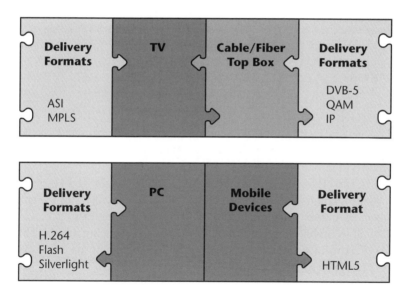

FIGURE 6-5

Figure 6-5 shows each of the protocols and the platforms they service. There is no longer a simple term called "conventional broadcast"; DTV changed that. Over-the-Air is now a multi-channel delivery system with integrated mobile services, with the studio to transmitter link (STL) as an IP stream.

The broadcast center is now an origination center and delivers to cable, satellite, and IPTV operators with DVB S, S2-ASI, DVB T, and T1-IP. The cable, IPTV, or satellite operator aggregates the channels for delivery using QAM via a set top box that also provides Over-the-Top services. The broadcast center delivers to CDNs for broadband, web, and mobile services with an interactive return path for offering additional services and enabling transactions. There are a number of transport protocols and formats based on where the stream is going.

Destination	Platform	Protocol
	Over-the-Air/Cable	QAM
TV and Set Top Box	Satellite	DVB-S, DVB-S2, DVB-T
	IPTV	MPEGTS/RTMP/HTTP Dynamic Streaming
iPhone, iPad, iPod	Apple iOS	HTTP Live Stream
Android, Blackberry, Other Mobile	3GPP	RTSP/RTP
Windows 7 and 8 Mobile	Silverlight	HTTP Smooth Streaming

FIGURE 6-6

This changing landscape for distribution further re-enforces standardizing the media formats and management in the production and library processes, allowing transcoding to resolve the changes in delivery requirements.

SMPTE is working on new protocols like MPEG7 and 21, and H.265. These new protocols will impact compression, overall content quality, and the amount of embedded metadata that can be carried within the file and/or stream.

The online, mobile, and On-Demand landscape is changing rapidly. There is a large effort to move away from royalty and license-based codecs (e.g. MPEG H.2xx) to an open source codec like VP8.

One thing that can easily be said is that there will be a constantly changing landscape in codecs, formats, and protocols. Staying on top will be an ongoing challenge.

In Figure 6-7, we see how from the origination center a single delivery for each format goes to a CDN, where it is replicated for final distribution to end clients. The client requests the content, the request is managed by the edge request router and metadata validates the user. The request moves into the distribution system, where it is confirmed, and the program is delivered. There are authentication keys

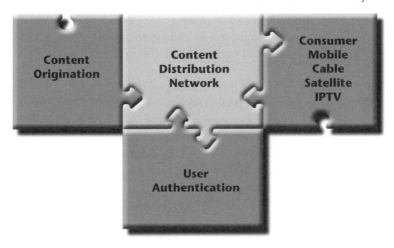

FIGURE 6-7

in a subscriber-based network that are transmitted back to the end user for validation before the content is released. The CDN is similar to an Over-the-Air transmitter in that its primary responsibility is to deliver the content to the end user device. In digital platform delivery, this takes on an interesting perspective. Back at the broadcast center, the content producer needs to format the programming for compatibility with all media players and browsers. The CDN needs to automatically detect which player and how much bandwidth the end user device accessing the content has and then adjust dynamically for all browsers to ensure the delivery is seamless and immediate.

Cloud Services

The cloud has become a prominent player in technology for all aspects of business and is rapidly moving into production and broadcast operations.

First, let's look at what the cloud really is. Until recently, the Internet was referred to as "the cloud." While that hasn't changed, the terms cloud and cloud service have expanded to mean remote datacenters that host many of the applications and services that the IT department and now the broadcast IT department were originally designed to support.

IT departments have been using virtual servers to obtain some efficiency in operations. This is done by using a single high-power multi-core piece of hardware and running multiple operation systems with multiple applications on the same hardware.

By definition and in practical application, there is a difference between virtualizing servers and moving services into the cloud. The most common use of the cloud is online storage through services such as Amazon EC2, Box.net, Dropbox, and Google Drive. However, the CDN (Akamai, Limelight, Kaltura) is also a cloud service. There are a number of broadcast and production products that are now being offered either in the cloud or as cloud services.

Cloud services are defined by three primary services:

Infrastructure as a Service (IaaS)—This is when a provider sets up the servers, network, and storage as an outsource, but the user is responsible for all configurations and management.

Platform as a Service (PaaS)—This is more common when the architecture is in place and the user loads their applications and can scale on demand. Amazon EC2 is an example of PaaS.

Software as a Service (SaaS)—Salesforce.com put SaaS on the map. In the media production industry, Chyron's Axis and Encoder.com are good examples. Adobe has taken this one step further with Adobe Anywhere and no longer offers their products as separate applications but only as a subscription service interconnecting their full bouquet of products. SaaS is good when an application can service a large user base from remote locations and offer collaboration and shared files without the need to host and manage the application on servers in the facility. The Subscription as a Service model opens both new opportunities and challenges.

Figure 6-8 helps clarify the differences between each of these offerings. In the packaged hardware and software, everything is on site in the broadcast facility and

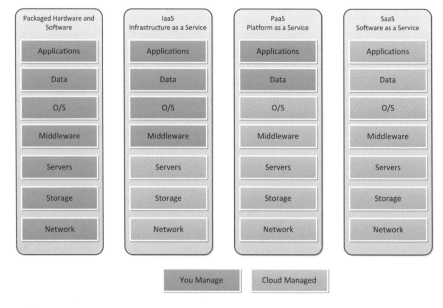

FIGURE 6-8

managed by engineering. In the Infrastructure as a Service (IaaS) model, as things are moved to the cloud, it provides the hardware for network storage and servers. The broadcast engineer is responsible for the operating system, middleware, application, and all data.

If it is run as Platform as a Service (PaaS), in addition to the servers, storage, and network, the cloud provider also manages operating systems and middleware. The broadcast engineer is responsible for any applications and managing data.

And last but not least, with Software as a Service (SaaS), the cloud provider manages everything—and provides it as a whole service.

There is also public cloud and private cloud. The public cloud is what most people are familiar with through Amazon, Google, Microsoft Azure, and iCloud. The public cloud is almost like a utility, where users can buy as much or as little processor or storage as they need and scale on demand. In the public cloud there can be multiple users on a single server. The private cloud services a single client on dedicated servers. As more capacity is requested resources are allocated but dedicated. The private cloud is offered through a different set of service providers, although Amazon offers both with its AWS EC2 service. A private cloud uses an offsite datacenter as a platform or host for applications. Most of the new products appearing as cloud services are running as Software as a Service (SaaS).

Cloud Services in Broadcast

In using cloud services for contribution, field producers can upload content to the cloud, and the broadcast center can download it without needing to create a direct connection. There are a number of initiatives to look at content distribution in the cloud where a programmer hosts their content with a cloud provider and their distribution channels can browse and access from there. There are other services where proxies are in the cloud for review and approval. As automation systems support the Channel in a Box and with CentralCasting, the automation systems can be cloud based.

And as mentioned earlier, the Adobe Anywhere product is a hybrid of an application local on the end user's machine that only stays activated by subscription to the cloud in real-time. It also keeps the core application in the cloud service, updating the local system each time the user logs on.

The most obvious use of cloud services is media delivery. The broadcast origination center delivers the content to a distribution network in the cloud. The distribution network has the necessary applications to host and replicate the content as needed, manage user authentication, and integrate interactive services.

Archiving and media management are other practical uses of the cloud. Archiving typically is a combination of a digital tape library or low-cost spinning disks. Both are a large capital investment plus have recurring maintenance costs. Placing the archive in the cloud reduces the footprint, the need for a protected environment, and overhead costs like electrical and mechanical services. It allows the archive to be accessed from anywhere and provides a level of disaster recovery.

Media management is another use of the cloud, as noted earlier, proxy storage increases faster than high availability high resolution storage. Placing proxy and metadata in the cloud makes the inventory accessible from anywhere, the tools enable a user to create an edit decision list (EDL) by browsing the proxies and instructions can be sent to a rendering agent in the facility to put together the final version for distribution. The orchestration or automation application manages the rendering and distribution.

The CDN also manages any transaction or subscription services that are available. Online media services such as Hulu, Vudu, Amazon, YouTube, Netflix, and iTunes are now considered cloud services. They use Akamai, Kaltura, Limelight, Brightcove, AppleTV, and others as their online video distribution network.

More and more broadcast and production products are simply software applications and many production and broadcast manufacturers are developing their products to operate in the cloud. What impact does this have on facility planning and design?

What cloud services work best for broadcast? And what needs to be considered when assessing the viability of these cloud services?

- Contribution
- Craft and Production
- Management
- Distribution
- Automation

Case Study 1

In the US, the Public Broadcasting Service (PBS) has driven a number of initiatives to consolidate program and origination plus master control operations to centralized and cloud services. The Corporation for Public Broadcasting, the parent of the Public Broadcasting is the central repository of programming from the member stations plus commissioning a lot of content. As an early adopter of file-based workflow, PBS still sends tape to all the member stations for airing. PBS is looking to create a more efficient system and is looking at the cloud to host all the content and have all the member stations access it. In addition, a separate initiative is to have each of the stations' traffic schedules integrated and let a cloud provider send the individual station program schedule to their transmitter, effectively running a multi-channel master control service in the cloud. This would let the member stations reduce operations for program origination and focus on production.

Case Study 2

There are vendors and service providers offering cloud solutions as central delivery of content as the next generation of centralcasting. In a different service model than PBS, these offerings are to enable commercial stations to reduce their overhead by using cloud services for all the master control functions, i.e. program origination and commercial integration, play-out to multiple platforms in multiple format and traffic integration

Cloud Decisions

What are the decision points to using the cloud? What's the most practical and the most effective way to use cloud services?

One of the core advantages of the cloud is scaling. Scaling in the cloud is on demand and supports stable growth. One advantage to using a cloud service is that when there is a peak demand, rather than building out to support it for one time and then having too much capacity when the demand subsides, you use the cloud as needed and scale back when the demand is lower. Cloud works! In the cloud, you can add more capacity, pay for it, and turn it off when the demand subsides.

A few of the first decisions when moving to the cloud are:

- Private or public cloud?
- Rent or build?

Contribution—there are a number of new initiatives beginning to offer cloud-based services using open Internet as the carrier for contribution. Obviously, bandwidth, security, and reliability are major factors here.

Craft and media management become more difficult or have limitations when considering cloud. What applications are most suited to the cloud? Is there a real benefit? Can cloud applications meet production requirements?

Distribution to web, portable, and mobile devices has always been a cloud service; it just didn't have the name yet. Akamai, Brightcove, and Limelight have always worked in the cloud. YouTube is a cloud service.

The following list shows in detail the different aspects of public vs. private cloud services and some of the roles for which it is worth using cloud services.

Public Providers

- Infrastructure as Service (IaaS)
    - Amazon EC2
    - Microsoft
    - Rackspace
    - IBM
    - HP
- Platform as a Service (PaaS)
    - Google App Engine
    - Microsoft Azura

- Amazon AWS
- Red Hat Openshift
- Heroku
- Software as a Service (SasS)
  - Chyron (Axis)
  - Encoding.com
  - Kaltura
  - Adobe Anywhere

Public vs. Private Cloud

Public is just that, accessed via open Internet and offering infrastructure, platform, and software as a service. These are managed and hosted services in remote data-centers that can be rented by processor, storage, and bandwidth. Multiple users share servers.

In the public arena for infrastructure and platform, Amazon, Google, and Microsoft have offers. While this is a business-to-business service, it still uses their core datacenters, servers, and storage.

Public cloud is a full outsource: renting what is needed for the services required. One of the logical choices for cloud services is setting up media-rich websites for On-Demand and streaming. Developing software for mobile apps is another good use, especially since the cloud providers offer load testing. From a security perspective, developing in the cloud avoids any danger of compromising live production tools. Cloud storage and file access from all devices is becoming more popular as an alternative to VPN and Citrix.

In broadcast and production, there are a number of vendors offering services in the cloud in the Software as a Service model. Chyron's Axis, Encoding.com are changing the thought process in planning and design when looking at infrastructure planning.

Private Build vs. Outsource

Build

- Provide access for remote user applications
- Content and metadata accessible
- Large capital investment
- Operating overhead
  - Maintenance
  - Power and Cooling
  - Upgrades
  - Scalability

Outsource

- Self-managed hosted servers and storage
- Self-managed hosted applications
- No capital costs
- No overhead costs

The private cloud comes in two flavors. One is to build or leverage your own data-center, creating a service model where users don't have any applications on their desktop and must therefore access tools via browser or remote connection.

In this model, whether on- or off-site, you are still building and maintaining a datacenter. It changes the infrastructure build-out and concentrates the servers and applications to the datacenter.

The other option is to use one of the cloud providers to build-out on their premise a dedicated cloud environment where they host and manage all the servers, storage, and applications. This philosophy and using the IaaS and PaaS models will relieve the need to build and manage a datacenter. If more capacity is needed, the cloud provider can provide that, either on demand or for permanent growth.

Back at the broadcast center, having adopted the cloud, instead of designing a server farm and large storage environment, it's all about network and bandwidth.

Not all production and broadcast functions are applicable to cloud, but this is changing rapidly. These are some of the areas where cloud can be applied:

- Remote production contribution
  - Truck
  - Field
  - Studio
- Outsourced production
  - Post production services
  - Graphics
- Automation
- Central casting
- Independent production
- Asset management and metadata log-in
- Search and browse
- Dailies review and approvals

One area that is seeing potential in the cloud is contribution. Where the open Internet has security and consistent bandwidth issues, the cloud services are limited by first- and last-mile bandwidth (last mile . . . nothing's changed here). There are a number of companies offering transport over open Internet with clever algorithms that optimize and accelerate transport. The broadcast engineer needs to plan his infrastructure to support this. Security, firewalls, and bandwidth are just some of the considerations.

Cloud-based craft services where graphic elements are created can be hosted in the cloud and not fully assembled until they are downloaded and rendered where the program is being produced. Cloud storage is also handy for large file transfers.

Using the cloud to edit might look like this: a producer or editor opens the asset management system (via browser) and finds the clips they need as proxies. Using the asset manager, they create the EDL. Once the EDL is complete, it's sent to a rendering engine, which does the conform and renders to the final product.

The cloud offers producers a way to review production dailies in proxy form. By making some or all of the asset manager accessible in the cloud, media logging can be done from anywhere by watching a proxy version of the production and entering metadata. Once the asset manager is accessible, any user can search and browse the library from remote locations.

Distribution has been in the cloud—specifically to online, broadband, and mobile—for a while. Content distribution networks like Akamai and Limelight have been cloud services since before it was in vogue. Popular video services like YouTube, Hulu, Vudu, and Netflix are all cloud services that distribute media.

The cloud is ideal for On-Demand media delivery and is used extensively to authenticate users for online delivered media that is subscriber controlled. The cloud service manages the decryption of protected content delivered to a PC, tablet, or mobile device once an authorization key has been confirmed.

The cloud connects the viewer to an engaging multi-screen experience, tying together social networks, private communities, and enriched program experiences as a companion to Over-the-Air, cable, satellite, and IPTV large-screen viewing.

The cloud brings another piece to our puzzle and that piece is helping to define what the whole picture of the IP- and file-based architecture will look like.

seven

FACILITY PLANNING AND DESIGN

Planning and design of the IP- and file-based infrastructure is as much about bricks and mortar as it is about technology. Other pieces to this puzzle are the many facets in the technology architecture and infrastructure. But the puzzle would not be complete without the pieces for physical facility and facility design.

These pieces include the changes in physical design for studios, control rooms, and equipment rooms. They are changes in the mechanical and electrical designs driven by the installation requirements of IP technology. How does the integration of enterprise and broadcast networks affect the placement of technology? How does it impact cable design and management?

There are a few core elements critical to the physical side of planning and designing a facility. They include:

- Space Planning
    - Space Adjacencies
    - Studios
    - Control Rooms
    - Media Operations (logging, ingest)
    - Program Origination Center (Master Control)
    - Craft Editing and Graphics
    - Core Equipment Room
    - Network Operation Center

- IT Infrastructure
  - Network Switch Locations (MDF/IDF)
  - Demark Rooms for Carrier Services
  - Enterprise and Broadcast LAN Locations
- Electrical
  - Critical Power
  - Protected Power
  - Power Distribution
  - Cable Pathways
- Mechanical
  - Air Flow and Distribution
  - Temperature and Humidity

Operating Costs—Space/Power/HVAC

The amount of physical space also factors into the total cost of ownership, including basic overhead costs such as space, power, and mechanical systems. It is a reasonable assumption that if there is less space in an IP-based architecture, it's also possible that the overhead is lower, in terms of potential power cost-savings. The amount of mechanical support systems needed in control and support spaces is also reduced significantly. IP-centric control surfaces for production devices use less power and are connected over Ethernet to a server in the equipment center. Video servers in comparison to video tape machines typically consume less power and the control rooms have less heat-producing equipment. This translates into needing less HVAC equipment and lowering operational costs.

In terms of mechanical systems, servers, storage, and networks tend to be consolidated into the equipment center, with only control surfaces in the craft suites. There are different HVAC design considerations since the servers, network gear, and data storage have a lower tolerance for heat. It is more critical to maintain the right temperature, but the overall amount of cooling required is a little less. IP equipment does not handle power outages well and is very sensitive to high temperatures. It is common to use UPS (battery backup) on all current production equipment, even without a generator to allow them to be shut down in an orderly fashion in the event of a power disruption. There are still heat- and power-related issues.

Modest control surfaces and new flat-panel display devices put out less heat. Workspaces for editing, graphics, studio and master control rooms are also different. These workstations don't occupy as much space, require as much power, or generate as much heat as VTRs, audio and video support, switchers, mixing consoles, editor and support equipment.

The interdependency between spaces has changed as well. With the data center now serving as the media backbone, media is now more readily available over the network.

In the IP facility, there are considerable changes to all control room designs. There are fewer control surfaces because of the features and functionality built into the equipment changing the layout of consoles. There is a huge difference in what an edit room used to look like in a tape-based world compared to what one looks like now in a file-based world. Multi-viewer monitor walls are dynamically configurable, with things like clocks, Tally, source ID, and audio monitoring integrated into the display. An operator can see and control all the devices, applications, and access to media in a smaller, easily managed space.

Linear   **File-Based**

FIGURE 7-1

The origination center, or master control room, now looks a lot more like a network operations center, while the main equipment room looks a lot more like a data center.

FIGURE 7-2

Other puzzle pieces are servers, storage, and networks. Fiber optic cable is probably the real backbone of IP architecture, with categories 5e and 6e as the copper cables used to reach end points. Fiber and Cat6 are handled differently than coax, multi-pair audio, and multi-conductor control cable.

Most server-based media technology and dedicated devices have at least two (2) network ports: one for media and the other for control or management. Some have more for redundancy, they may go to the same switch, but they are configured on different VLANs. Even simple things like cable management are different. Category 5e and 6e (and coming soon, 8e) cable does not like to be squished or compressed! This means that instead of nylon cable ties, hook and loop (Velcro) is used.

With IP-based technologies there is very little equipment that has front panel controls anymore, user and admin access is all done through a software interface or a display with keyboard and mouse control. Servers are deeper than legacy broadcast equipment, so allowing for service and cabling in the racks is important. Network switches have their ports in the front, so cable management changes. It is important to make sure there is enough space to install and remove the servers, so this changes the amount of space needed in front of the rack. This translates into each rack occupying more space, therefore reducing the number of racks in the equipment room. Equipment density in racks has changed where some servers can be stacked tightly however some have side vents for cooling which impacts rack width and spacing. At the same time they occupy fewer rack spaces than single purpose proprietary production devices (i.e. tape machines).

The equipment room now more closely resembles a data center. Master control resembles a network operation center (NOC). As a result of this, space, power, and environmental conditioning design needs to be modified to meet the requirements of servers, storage, and switches.

IP- and file-based technology has a significant impact on the physical design— that is to say, the bricks and mortar of the facility.

Space Planning—Room layouts are different; the number of operating positions has changed, and spatial adjacencies, cable pathways, and access are all different design considerations.

Mechanical and Electrical—Servers, storage, and switches have different power and environmental needs than tape-based baseband. In some ways, the devices are more forgiving; in many ways, much less.

Connectivity and Telecommunications—These are almost the same thing in IP. While telephone (voice) and Internet come through a different service, they conceivably come from the same service providers—and telephone and intercom are now VoIP. There are still dedicated providers that specialize in moving media; however, the larger communications carriers have moved into this space as well. The IP STL is a fiber link over a common carrier network. And there are a number of new vendors using the open Internet with their own algorithms to make the transport more efficient. Now it's all about bandwidth.

Network design has a core switch in the main equipment room and satellite switches closer to the end device in Intermediate Distribution Frame (IDF) rooms. There are more pathways for fiber and cable to service them. The entry point into the premises for high bandwidth fiber services needs power and environment conditioning.

Studios—In designing studios, size is still first and foremost, and is decided by the kind of production. That being said, the lighting instruments use fluorescent bulbs (CFLs) and LEDs that are cooler, use less power, and emit less heat into the space. They are controlled over IP.

Virtual sets and large monitor walls that are used as backdrops need less space than physical sets. Studio cameras and lenses are smaller, and robotic camera pedestals can occupy a smaller footprint. Higher camera sensitivity allows the lighting designer to use fewer instruments while still creating a dramatic visual effect. One of the crucial issues no different than before is when using robotic camera pedestals the floor must be level and smooth.

Studio Control and Support Rooms—One major change in the studio control room design is that it can now be fully automated, manually operated, or a hybrid. The physical size of the control room now depends more on the number of production personnel positions than the number of operating positions. The production switcher has digital effects, image store, and clip player fully integrated to one control surface instead of separate controllers. When using robotic cameras, a touch panel can be programmed with a shot list that includes camera moves. The touch panel can sit on an articulated arm in front of the technical director (TD) and over the production switcher. A single screen and keyboard handles all playback control on a play-out server located in the equipment room. The video operator has a master console for the camera control units (CCUs), and the camera connection is a hybrid of copper and fiber. The operator has a multi-screen display that, in addition to the camera shot, shows the waveforms used to balance the cameras (with memories). The audio engineer has a control surface that resembles a mixing console, however, the server frame where all the audio sources connect resides in the equipment room. The operator configures the mixing console with a keyboard, mouse, and monitor.

Master Control—Program Origination

The master control room is more of a multi-channel, multi-platform program origination center. All program providers deliver multiple channels over multiple delivery platforms. It's the master control center that manages program distribution and quality control. Even single channel television stations are now multi-platform delivery networks. Over-the-Air can have up to six (6) full channels of SDI, or three (3) full HD channels and fourteen (14) mobile DTV channels. All program origination facilities support broadband, On-Demand, and mobile and these are all "channels" that can have their own play list, play-out, and need monitoring. The program origination center has multi-view displays with audio metering for each display. There are high-resolution monitors for quality control, and keyboards, mice, and screens to control all play-out and management. The play-out servers are located in the equipment room, so the physical master control room only needs enough space for operators, keyboards, and flat panels.

New Spaces—Media management and ingest manager control rooms need to be designed into the facility. The operator positions are workstations with multi-channel displays that control the movement of the media. The actual space requirement is not large and there is very little heat and power necessary. It has similar physical requirements to an edit room.

Post Production or Craft—This is editing and graphics. Here, there have been significant changes in the physical requirements. These are all workstations connected to servers or dedicated machines running high performance applications. The consoles have one to three monitors, a keyboard, and tracking device. There may be a separate networked computer for enterprise. There are no longer dedicated controllers and tape machines. Even for the most complicated edit or graphic production, it's still a single workstation. The workstations do not draw a lot of power and as result do not generate much heat. The physical size of the rooms has decreased, as have power and heat loads.

Core Equipment Room—This room hosts more systems than it did before IP. Now that most production is server-based, the servers are located in the core equipment room. In the core equipment room, there are still IP and non-IP systems. When designing the core equipment room, consider cable density and equipment density for power and heat distribution. What are the touch points between SDI and IP? How should the equipment be laid out in the most efficient way for signal flow and cable management?

The QC position has changed; there are still test and measurement monitors with SDI router control to assure the integrity of SDI and AES audio and video. Now in addition to SDI test and measurement, there is a rack-mounted keyboard, tracking device, and monitor that can access all the servers and QC applications for monitoring files and streams.

Applications, servers, switches, and storage don't necessarily need the kind of clean power SDI equipment does; however, they do need protected power. Even devices not considered critical still need adequate time to shut down in the event of a power disruption.

Server, application, and storage crashes can be harmful. It is more complicated to shut down an application, disk array, and servers than it is to turn off a tape deck. Servers, switches, and storage can run at different voltages, so power distribution changes. Servers, switches, and storage will shut down on high heat.

Traditionally, facilities were designed around the central equipment centers as the core, with studios and control and edit rooms close by. Tape machines for record and playback needed an operator position with monitoring and measurement. There was a lot of cable traffic moving between rooms, and there were dedicated control heads for each production device.

The move to an IP infrastructure changes that.

- Facility Infrastructure—building
  - Core space planning
    - Adjacencies
    - Space allocation
    - Equipment center
  - Ergonomics
  - MEP systems
  - Telecommunications
- Wire management and cable planning

Similar to the construct of an enterprise network, operators and workstations can be located anywhere as long as the network topology is set up to support it. There are still obvious reasons to keep core operations together, some having more to do with physical plant design.

It's easier to construct acoustically sensitive spaces near each other and away from non-technical operations. However, media management and even craft edit and graphics can be closer to the production team instead of closer to the equipment center.

From an architectural (building) perspective, it's still more cost effective to build an "acoustic envelope" of spaces rather than disbursing them around the plant. It also makes more sense mechanically and electrically.

But with studios, even those needing a lot of space (e.g. for an audience), they can still be more efficient. On-camera monitors and displays take up less space, require less power, and produce less heat.

In studio lighting, the alternatives to incandescent, halogen, and tungsten continue to improve, having a considerable impact on heat and power. In a comparison between an 860W Tungsten light and an LED alternative, the LED draws 36 watts of power for the same amount of light. LEDs can last for up to 100,000 hours and require approximately 75% less electricity and generate up to 80% less heat compared to conventional lighting. This has an impact on both the cost to build and the cost to operate.

Take a look at the whole impact: the dimming system can be smaller, and so can the distribution wiring, resulting in less power service, lower electrical bills, less mechanical environment, and, if the lighting is on generator, less load when sizing.

Edit rooms need less space; even the beefiest workstations require less power and cooling than a rack full of tape machines, DAs and audio/video processors. These are now all software plug-ins to the edit system. Monitoring is a single screen driven by a multi-viewer.

From the equipment center, distribution uses high bandwidth fiber to interconnect switches. Instead of individual cable, high capacity fiber connects from the core switch in the equipment room to satellite switches in the production control or edit rooms, carrying audio, video, control, and communications. The end devices connect over Cat5e or 6e cables to the local switch, and the switch manages whether the end device gets media or other control or management data. A single cable or fiber feeds the multi-view display and remote access over IP allows operators to reconfigure the display wall. There is a reduction in cabling and all the other associated costs in running large bundles of different cable types.

Another benefit to an IP infrastructure is that when a new device is added, it's a port on the network and there is no need to run additional cables from the equipment center. It is a local connection within the control room, and the engineer can access the core and local switch to add the new device to the system.

In the equipment center, there are changes in rack layout. Servers can be 36" deep, meaning racks are now 36"–42" deep. When racks face each other, there needs to be a minimum of 48" (4') to allow equipment to slide out on rails. Therefore, racks need no less than 4' of front clearance. And since network switches tend to front load and device connections are still rear, it is becoming common practice to rear face switches or recess mount them so that the cables and cable management does not protrude from the front. These changes also affect the rear clearances as well. Where before a minimum of 2 feet was allowable, now 3 feet is becoming more of the standard. This means that the overall footprint of a single rack is now approximately $10' \times 2'$, or 20 ft^2. This has a huge impact on space planning.

Ergonomics has always been a consideration in control room design. It's all about how many things one person needs to look at and how many things they need to operate without straining their necks to see or constantly change their position to operate. The capabilities of production devices functionality is now concentrated into fewer devices. Images in the multi-viewer display are actually closer together because there a fewer bezels and no mullions from equipment racks. Also, the multi-viewer displays are easily programmable and can simultaneously show audio metering, source ID, and tally for each image.

Edit and graphics stations can have smaller consoles, with monitors mounted on articulated arms that keep them in the main field of vision. The operator uses a keyboard and tracking device instead of multiple control heads. In the file-based and IP-production infrastructure, media is accessed from folders on a storage system instead of routing a tape machine to an input on a device. This changes the type of repetitive physical movements in the production rooms.

One of the largest costs and complicated design considerations in building and operating a facility is the mechanical and electrical systems. In planning and design of the electrical plant, achieving clean power and good ground have always been the bane of all engineers. It has kept the HumBucker (Allen Avionics) people in business for a long time. While clean power and good ground practices are still important, there are a lot of changes that started with SDI and have continued to evolve ever since.

Theoretically, there is no ground noise in digital and switching power supplies that can introduce interference in the signal. This is mostly true. Look at what has changed in electrical and mechanical design for IP.

Servers, spinning disks, solid state storage, and network switches depend a lot on stable and consistent power and temperature. There is a slight difference in humidity specifications for IP technologies vs. SDI, but not much.

This may sound like a primer, but consider the difference between turning on a tape machine vs. starting an encoder. With a tape machine, you would flip the "ON" button while confirming audio and video is present, this includes entire signal path of routing, distribution, frame synchronizers, and other terminal equipment.

With an encoder, you power on the server and see that the BIOS is OK. Then, the OS starts up and needs to load, next up are the services for the hardware and databases, while establishing the network connections. Next up are the services that support the applications that handle the encoding process. After all that the applications that should auto-load need to start. Are there any manual applications that need initiation? Nothing corrupt? All good?

Now, when power fails with a VTR, first the power switch is turned off so that when power restores, any surge doesn't impact the machine. If there is a tape in the machine, it stays there until power is restored, when it can be examined and is hopefully fine, but if there's a problem or it's jammed, it's only that one single tape.

On the server side, when the power goes out, all running applications crash, the network shuts down, and disks stop spinning while the heads are engaged. If this is in the middle of a process, that process could corrupt both the media, disks, and the application. There may be an impact on the operating system. Many servers do not have a physical power switch, so a restore with a surge could affect the power supply. Abrupt power failures can damage hard drives, as the disks are spinning at high speeds and a power outage could cause a head crash. Servers like to be turned off carefully so that processes and processors can stop in an orderly fashion. If there were files open or in transit, there is also the potential for file corruption. This is a good time to review the need for that archive and backup plan.

Now compare planning the electrical system. The IP architecture overall uses less power, yet at the same time, the design of the electrical systems is different. Servers, storage, and switches typically have dual power supplies wanting to be on different circuits and possibly different legs. They can also operate at different voltages; for example, instead of 110V single-phase, they can use 220V three-phase. This reduces the physical size of the power supply and allows them to operate at lower current loads. The design of the power service and distribution are also impacted. Servers, storage, and switches prefer not to be shutdown abruptly; they have shutdown sequences. Not all systems need to be on full UPS and generator back up; however, having enough UPS to give the servers, disks, and switches enough time to shut down gracefully is highly recommended. This only takes about 5–7 minutes, but it is an important 5–7 minutes. Many of these devices have a lights-out feature that ties into the UPS. In the event of a power outage, the lights-out feature tells these devices to go into an automated shutdown sequence.

If there is UPS and generator backup, this adds an additional layer to power distribution. There are three types of power; street power which is the service provided by the electric company, protected power which is UPS (Uninterruptible Power Supply) and critical power which is UPS with generator. There is actually one more layer which is generator only with UPS so there is a disruption in service until the generator starts, and generators are known to surge which can damage technology power supplies. It's good practice to have critical systems on UPS and generator power, and non-critical systems on UPS only.

In control spaces, there are fewer devices, which draw less power. Another benefit to the reduction in power loads is that the mechanical loads are less, reducing the size of mechanical systems—which also has an impact on the electrical design.

Speaking of mechanical systems, IP devices are being designed to run at higher temperatures, so even if they have an equal (if not greater) sensitivity to heat, the amount of cooling needed is reduced. And although humidity is still important, they are a little more tolerant (but condensation on disks is still frowned upon).

Servers and disks generate a fair amount of heat. Similar to broadcast, they typically have fans in the rear that draw cool air through the device, expelling heat to the rear. So in planning the equipment room, or controls rooms for that matter, it is important to have cold and hot aisles that manage airflow. The cold air (supply) is ducted from overhead in front of the rack and the return is also ducted and draws from the rear of the racks creating the correct air flow and managing the heat.

Because servers tend to be stacked tightly into racks, there is a lot of heat generated in a relatively tight space. Keeping air moving and getting the heat away from the devices is critical. Server farms are being planned with hot and cold aisles to

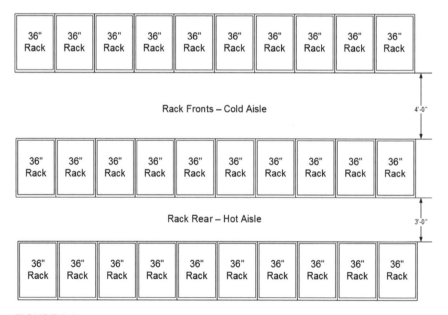

FIGURE 7-3

manage the air flow. With IP technology, the actual loads tend to be consistent, so good planning with some allowance for growth will provide good results without surprises.

Figure 7.3 is a quick sketch to demonstrate the new distances and allowances in the equipment room.

The fronts are the cold aisle, with the cold supply air in the ceiling—cold air falls, and as mentioned earlier, with servers being deeper, there needs to be enough separation to pull a server out for servicing or replacing.

In the hot air aisle, the return air is also ceiling mounted, drawing the heat away from the racks and devices.

Case Study

On a major facility build out project, the organization engaged an architect and MEP team that used older data center design philosophies. Earlier design for data centers put positive pressure cold air in the raised floor and hot air pulled from the top. In this instance the cold air was in the floor and the return (hot) air was open in the room. This put undue strain on the equipment and created an overheating situation where equipment was shutting down. To remedy the situation as best as possible, air blocks were put in the bottom of each rack, perforated tiles were strategically placed in the floor to maintain enough pressure to get the cool air out of the floor enough for the equipment to draw it in before the floor mounted units pulled the return air before it could reach the equipment. An expensive design that didn't work.

Space

Building on the space conversation, control rooms can be more efficient in layout and need less space for control surfaces. Now that we have touched on automation, we can discuss a studio in a box and station in a box. These have fully automated studio control and origination control. All the broadcast and production devices are either software-based or dedicated hardware controlled by software. These are single operator systems managing programmed sequences. In the studio, automation includes camera positioning and the manipulation of virtual sets. Control rooms that are fully automated don't require the same amount of support people. Editors now preview and assemble their EDLs from the asset manager and then use the full production editor to conform and render with all the effects and elements.

Workstations for graphics have been around for a while, possessing elements available in common storage or cloud. Searching is easier and improved in the file-based workflow. And we just saw the impact in the equipment room.

Telecommunications, Voice, and Data—Telecommunication is the backbone to the IP architecture. The bandwidth and transport has been examined now it's time to add in the rest of the telecommunication services. Telephone services now include voice over IP (VoIP). On the production side, intercom systems are also VoIP. Telephone is typically handled in the enterprise design. But since production intercom, telephone hybrids, and other former discreet phone services are now IP, these need to be included for a complete design. Production and contribution use audio and video conferencing, Skype, and other services that integrate into the production.

It's a good idea to include wire management and cable planning in the physical plant discussion. The broadcast industry actually does not have published standards for cabling and cable management, there are however industry accepted practices. In the IT world, there are published standards for structured cabling (EIA568B). Some of this has been adapted to broadcast installations; however, there are differences—specifically in how horizontal and vertical cabling is handled.

FIGURE 7-4

It is certainly true that using a single cable type that carries multiple signals reduces the number of cables. Also, since a lot of the signals are aggregated in a local network switch and then sent over fiber to the core switch, cable management may be more sensible. Fiber tends to come in multiples by number of strands. Fiber is very thin, so a lot of strands don't take up much space. Conduits are fewer and smaller. What is important to fiber is bend radius, since it is glass and it can break. Cat 5e and 6e have standards to follow. However, when designing the cable plant, the troughs or conduit still need to consider the number of dedicated runs and allow for growth. Ethernet cable does have to stay away from electrical, typically crossing at 90 degrees.

Engineers know that all cabling is perfectly dressed and laced, at least for the first 15 minutes after the installation is complete. And then the changes and additions happen. Fiber and Ethernet cabling is dressed and harnessed. One small point of interest is that the telecom and data industry has standards for cabling, ANSI TIA/EIA 568B Structured Cable Standard that includes labeling, conduit sizing, bend radius, isolation, and management. The broadcast industry for all its standards and protocols does NOT have a cabling standard, only recommended industry practice. Now we can begin the next topic of cabling standards for Ethernet and fiber optics.

There are a number of organizations that produce and publish the various standards for the broadcast and production industries. SMPTE, IEEE, ISO, EIA, and ITU are the standard organizations for broadcast and production transport, signals interfaces, and processes. For analog video, the US and Europe both accepted the EIA RS170M and 250C as standards; in addition the ITU uses the BT470 designation.

In moving to digital, SMPTE and ITU came together to standardize SD/HD-SDI with SMPTE259, SMPTE 292, and ITU-R BT709.

ISO and IEC joined to form the MPEG working group, partnering with SMPTE and ITU and producing the myriad variations for the multitude of compression algorithms and protocols.

In the physical design of an IP facility, EIA is the standard bearer, with the well-established EIA 644 standard for data rates over twisted pair and EIA/TIA 568B as the current standard for IP cabling topology.

When running network cables, you will hear terms like "MDF" and "IDF"; these stand for main distribution frame (MDF) and intermediate distribution frame (IDF). The MDF is located in proximity to the core switch, and the IDFs are aggregation points of end devices typically with a satellite switch.

Fiber optic cable handles the high bandwidth interconnection between the core switches and satellite switches and resolves distance limitation problems. In addition to bandwidth limitations, twisted pair has distance limitations similar to that of coax.

Figure 7-5 shows the different bandwidth topologies that represent the interconnections between the core switch and sub switches. The core switch attaches to end devices over 1Gb/s and 10Gb/s connections, there is high bandwidth 40Gb/s connections to the satellite switches and 100gb/s between the cores switches. The high bandwidth connections are all on fiber optics. Typically from any switch to an end device is 1Gb/s over twisted pair. When fiber is used there are different types of fiber as well, there is single mode and multi-mode and the decision regarding which to use is based on distance and bandwidth capacity. The main difference between single mode and multi-mode is the optical wavelength and the distance the wavelength will travel. Multi-mode is typically used for shorter distances, which can vary based on how much bandwidth is needed. Single mode fiber has fewer limitations and can handle greater distances. Another advantage of single mode is the ability to use multiple wavelengths on a single fiber. This allows multiple signals to be multiplexed on a single fiber pair.

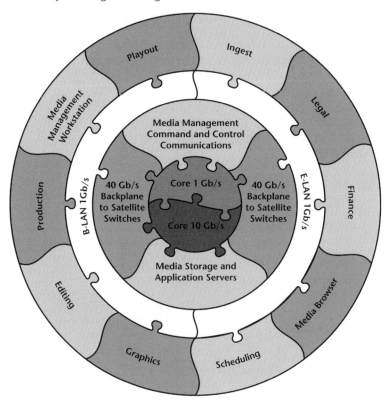

FIGURE 7-5

When calculating cable management, if there is enough space available, keeping cable types segregated will make them easier to manage. When choosing fiber, sleeves, or buffer tubes can come with a mixed set of single and multi-mode strands. These can be mixed if there is a combination of low bitrate and high bitrate traffic in the same path (e.g. video and automation control). Fiber optic cable is typically run through innerduct, a flexible conduit that complies with electrical codes and is easier to work with than rigid conduit. A fiber sleeve with twenty-four (24) strands is about the same diameter as a single coax cable, enabling more capacity while occupying considerably less physical space.

The IP- and file-based broadcast environment is based on an IP network infrastructure. There are different network requirements to support media. Some are the same as in the enterprise; however, most are different.

- Network switches are the new core components:
  - □ There are new forms of routing and distribution
  - □ VLANs segregate media from command and control

□   Media is a bandwidth and network hog
□   QoS is critical to media performance.

The actual network routers, switches, and firewalls are the new core in the media IP architecture. The topology of routing and distribution has changed. The router in IP design is very different from an audio/video router. The IP router is an access controller and it's the managed switch that controls signal movement. Signals move without the need for separate matrix controllers; IP addressing enables file and stream movement. The automation software triggers events to move files and streams. The IP switch uses trunking and access lists to control which IP paths interconnect and where separation is needed.

Network switches handle a broad variety of packet traffic. Managing that traffic is critical, and shaping the flow of data to ensure that there is no disruption in services means configuring the network with parameters and a priority structure—meaning determining which packets should have priority over others. This is known as Quality of Service (QoS), and is a critical configuration component of the network in media transport and management.

When planning the design and configuration, the engineer needs to consider the number of network segments, devices, distances, QoS, and latency requirements for each data type.

The most common devices in a large installation are layer 2 and layer 3 switches. The differences are in the features and functions between them. At the core, layer 2 switches provide high speed and low latency, but do require a router to set up multiple VLANs.

Layer 3 switches can be placed anywhere in the network because they handle high-performance LAN traffic and can cost-effectively replace routers.

Above, there are layer 3 switches in each operational area, with a firewall and router in between. Layer 3 network switches can handle the high bandwidth traffic and additionally have the control and access features needed in a VLAN infrastructure. The production and business groups both need Internet and VPN access. There is a firewall and router between internal networks and outside networks. All Internet services are outside the firewalls with controlled access. Even the streaming server sits outside the public firewall.

The thinking process for designing an IP architecture and infrastructure is a departure from that of the more traditional infrastructure.

- Infrastructure Technology
  □   IP Network vs. SDI
  □   IP Routers & Switches vs. AV Routers, DA, A/D, D/A
  □   Copper vs. Fiber, Twisted Pair vs. Coax
  □   Single Mode and Multi-mode
  □   Standards EIA/TIA vs. SMPTE/ITU-R
  □   Cable Pathways and Management
  □   Bandwidth Management
  □   Networks Segments

It's all about the network. Servers and storage can be physically located anywhere on premises or off-site. With enough bandwidth and the proper configuration, users will be unaware of where the devices actually are.

Network switches cannot completely replace an SDI router quite yet. In the IP architecture, a layer 3 switch and firewall are the new router, DA, A/D, and D/A. The designer needs to know when to transition from copper to fiber and when to use single mode vs. multi-mode. Good cable management means conforming to the EIA 568 wiring standard for telecommunications and computer networks.

When designing control and equipment rooms, today's computer workstations, small control surfaces, articulated equipment arms, remote devices, and multi-view displays make it easier and more comfortable for the operators and occupy less physical space.

- Operations Technology
  - Control Surfaces and Multi-viewers
  - KVM over IP Matrices
  - Robotics and Automation
  - Command and Control Embedded with the Stream
  - Database and Network Management

KVM over IP matrices are extenders that enable a single keyboard, tracking device (mouse), and screen to access and control a number of server applications from multiple user positions.

Robotics and automation, either fully automated or partially automated, create operational efficiencies. Network design is critical to ensure the automation commands trigger the appropriate event at the exact time.

The IP network carries the media and all the command and control data.

Metadata, database management, and network management are the core technologies in the IP architecture and the engineer's new responsibilities.

Let's review the physical plant and the considerations in design:

- Space Planning
- Electrical Distribution and Protection
- Mechanical Systems
- Telecommunications

Technical and production spaces can be more efficient. The networked infrastructure enables edit, graphic, and media management to be located anywhere in the facility. New lighting technologies reduce the power and cooling requirements and change studio power distribution.

Computer workstations for craft production and LCD panels for displays reduce power to the control rooms. In the equipment center there are changes in power distribution and load calculations. Servers, switches, and storage can operate on higher voltages, adding different electrical distribution requirements. IP devices need sufficient UPS battery to allow them to intelligently shut down in the event of a power outage. Even in instances where there is no generator, critical systems should be protected.

The reduction in power and heat generation in control spaces has a beneficial impact on the cooling loads. While servers, storage, and switches generate heat in the equipment center, there is less overall load. Changing the mechanical loads also has a beneficial impact on the power loads.

And since the entire IP architecture and infrastructure is based on a telecommunications topology, this is a crucial component of the IP facility. Telephone services are IP, and Internet access is more than just providing web services and email—it is a critical service to some of the delivery platforms. Telecommunications includes all the dedicated IP paths for STL and point-to-point delivery of content to multiple delivery platforms. On the production side, intercom uses VoIP and remote access to many of the services within the media management systems.

Monitoring and management in the multi-platform, multi-channel distribution ecosystem brings a new dimension when planning and designing the physical IP facility. Multi-viewers are embedded into router frames; they have audio metering and clock and countdown timers with under-monitor displays. They can monitor servers directly without needing a video card in the server.

Planning and design for an IP production center changes the way people have been working and interacting. The engineering team needs additional skills and knowledge. It wouldn't be a complete picture without looking at the physical plant and the changes there.

eight

The transition to IP in the broadcast and production industry is a game changer and a new puzzle to solve.

This book identifies many of the new puzzle pieces and explains where they may fit in the complete picture of the IP- and file-based infrastructure and architecture. The goal is to help the reader understand many of the requirements to meet these new challenges.

There are new considerations in the planning and design of an IP- and file-based broadcast and production facility. It's important to understand the changes the new technology brings and the impact it has on operations, workflow, and processes, including the physical design of the facility.

This book takes a big-picture perspective in examining the entire lifecycle of media and the technology architecture, workflows, and business process that make up the broadcast and production file-based and IP ecosystem. As the full picture of the puzzle takes shape it reveals the entire architecture inclusive of workflows, facility planning, and many of the changes in roles and responsibilities.

There is no single solution or "silver bullet" that addresses all the aspects of this change. From the beginning of broadcast television, there has been no single manufacturer or service provider that has had all the technology or solutions. Broadcast and production is an integrated environment and ecosystem of different technologies, services, and operations to create, produce, manage, and deliver programs. That has not changed.

Broadcast and production has been evolving over time. There have been many transitions and new technologies introduced. These were all in support of a single NTSC or PAL format for distribution. The technologies brought more features and capabilities with new standards. The industry went from analog to digital and then to high definition. The transition to IP- and file-based technologies converged in a perfect storm of change as new delivery platforms were introduced demanding more content in so many new formats.

The planning and design process begins with some core decisions. Content formats, file formats, bitrates, archive and retention policies all play a role in deciding which technologies are best suited for the facility. How much automation will run the facility? In the past these were not the primary considerations when planning or upgrading a facility.

The first pieces of the new puzzle introduced the changes in the capture or recording process. The creation process originates the first elements of metadata and where the transition from essence to asset begins. Metadata is one of the key elements in the IP- and file-based value chain and is the common thread throughout the ecosystem. The capture process is based on the formats and standards for streams and files. The acquisition process now includes in addition to cameras, computers, tablets, and phones as capturing devices. The recording is to a computer, server, hard disk, flash memory, solid state memory, optical, digital tape, and the cloud.

The introduction of new and multiple delivery platforms plus the addition of second and third screen interactive companion content has presented new demands on creation, production, and distribution. Creating program content for web, tablet, and phone is different than for a large-screen TV. All these digital platforms provide valuable information on what the consumer is interested in and their usage habits, using metadata. Metadata is stored and managed in databases that interact with other databases. Metadata is the command and control information that moves content throughout the IP infrastructure.

IP- and file-based broadcast and production is more than a DAM or a MAM. It is all the technologies that encompass the entire media lifecycle and value chain from acquisition and management to the transport of media. There is no single technology that addresses the entire lifecycle.

The IP- and file-based technology infrastructure is network, applications, servers, and storage. The network is the core technology and backbone that the entire media management architecture sits on. It is a multi-layer topology that needs to be carefully planned and designed. The network provides media transport, communications, command and control, management, and the integration between systems. There are different types of networks that integrate as a complete system, there are networks for transport (i.e. SONET, MPLS, and Fast Ethernet), infrastructure (TCP, UDP, UniCast, and Multicast) and storage (Fiber Channel, iSCSI, and UltraWide SCSI). The integration between the enterprise LAN and broadcast LAN depends on security policies and firewalls. VLANs segregate the different services on the network replacing discreet systems.

There is a tighter integration across the production, broadcast, and business units in every organization. This tighter integration is between the applications and databases each of the business units use, and that creates new workflows and processes.

Governance is the rules and policies that manage the media and metadata and control all the processes. These are the rules that tell the automation systems how to manage and move the media throughout the infrastructure.

The role of the broadcast engineer has expanded to network administrator, database manager, and applications admin. The diagnostics and tools to monitor, manage, and maintain an IP-centric environment are different than SDI. BitRate Error, Packet Loss, and Checksum are some of the new anomalies that can corrupt media. The design criteria of the IP cable plant is different, there is more fiber optic cable plus Category5E, 6E, and 8E copper and these have different parameters than Coax and multi-pair. The broadcast engineer and IT engineer work closely together to maintain the systems. Maintenance is sustainability, and the broadcast engineer needs to be more aware of the lifecycle of hardware and software.

The equipment room resembles a datacenter and the control room is now a workstation with large flat panel displays using multi-viewers, operated and managed by control surfaces, with keyboard, mouse, and display that access the servers and applications in the core equipment room using KVM extenders and matrices. There are different considerations for the architectural, electrical, and mechanical design. Some of these changes are more efficient and others are just different. The total cost of ownership of the technology includes annual service contracts and the overhead costs of the electrical and mechanical systems.

Transmission, contribution, and distribution technologies have changed. Cable, Satellite, and Over-the-Air are not the only distribution channels. There is Over-the-Top, broadband, and mobile. The STL to a transmitter is a fiber connection and the transport stream is ASI and MPEGTS. There are file accelerator service providers (i.e. Aspera and Signiant) that optimize the transport of files over IP networks for contribution as well as delivering programs to cable and satellite services providers. Files and streams are sent to content distribution networks (CDN) that also handle user authentication.

The cloud is changing workflows and posing questions on how much hardware infrastructure is practical to build in the facility that will have a limited lifecycle and have all the associated costs if the facility needs to scale to grow. The cloud offers scaling, access, and reduces overheads. Which processes will benefit by moving to the cloud? As the cost of bandwidth continues to drop, is it practical to move high value content to the cloud? Automation, content approval, and media management are all candidates for cloud solutions.

There are many more pieces to this complex puzzle. The broadcast industry continues to adjust and evolve, on the production side 1080P, 4K, and 8K will provide higher resolutions and quality and stay in SDI. However, if these are captured then it goes to file at very high bitrates and if it is live, there needs to be high bandwidth transport available. Multiplexing cameras onto single fiber paths could change the way remote production is supported. On the distribution side MPEG4

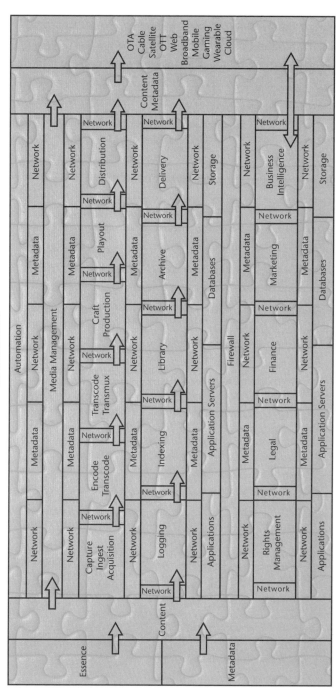

FIGURE 8-1

and HEVC are optimizing compression and with higher quality, allowing service providers to use less bandwidth to deliver programs.

The planning and design of the IP- and file-based broadcast center is a different knowledge base and skill set. The adoption of these new technologies and workflows is an entire change management program, and the topic of another book.

Glossary

Definitions

Digital Asset Management consists of management tasks and decisions surrounding the ingestion, annotation, cataloging, storage, retrieval, and distribution of digital assets. Digital photographs, animations, videos, and music are samples of media asset management

Digital Asset Management systems include computer software and/or hardware systems that aid in the process of digital asset management.

The term "Digital Asset Management" (DAM) also refers to the protocol for downloading, renaming, backing up, rating, grouping, archiving, optimizing, maintaining, thinning, and exporting files.

The term **"Media Asset Management"** (MAM) is a sub-category of "**Digital Asset Management**," mainly for audio, video, and other media content.

Metadata is the description of the asset and the description depth can vary depending on the needs of the system, designer, or user. Metadata can describe, but is not limited to, the description of: asset content (what is in the package?); the means of encoding/decoding (e.g. JPEG, tar, MPEG2); provenance (history to point of capture); ownership; rights of access; as well as many others.

Terminology

There are four core elements that are often identified as integral to digital management systems.

1. **Managing the Content**

   With a database operating in the background, a collection of assets is organized through some sort of user interface. User interactions with the assets are managed through this engine, whether requests are made to temporarily take content off-line or to ask for content "ingests" to the system or content "downloads" or "streaming" for consumption.

2. **Describing the Content**

   The actual media item is referred to as "essence." The item itself represents half of its value to an organization or consumers. The other half is captured in the well-formed descriptions associated with the item, whether the descriptions are about intellectual content, ownership, rights and use restrictions, or identifications of the forms and formats in which an item is available for review and playback. This is metadata, and it is captured in another engine driven by a database.

3. **Finding Content**

   With metadata in place, a search engine is engaged in order to allow consumers to find and display desired content. Searches can be conducted by keywords, database field values, virtual catalogs of items, or by hierarchically nested directories of content.

4. **Controlling Access**

   Often referred to as security logic, a layer of controls are present in order to generate types of user groups whose members have certain access and re-purposing privileges over content items. Access control is permissions control.

Glossary

Archive

Manages policies for retention in digital, on-line, near-line, and off-line storage. A well-managed archive enables efficient searchable access to inventory irrespective of location or format. Maintains effective accessibility to collections through cataloging, quality control of metadata and format, preservation, and conservation of digital and analog carriers of content.

Asset

Anything that has value to an organization can be considered an asset. An asset is a conceptual work consisting of physical and/or digital entities and associated metadata. Video, audio, photos, graphic art, and documents (i.e. Docs, Presentations, Spreadsheets, PDF) are all examples of assets. An asset may exist as a concept before an instantiation is rendered. For example, a scheduled event that has not yet occurred is a conceptual asset and the eventual recording of the event is an instantiation of that asset.

Asset Management

The term "Asset Management" when taken in isolation can be confused with Digital Asset Management. In general, it relates to the management of either physical objects, locations, or items of value and might include: computers, furniture, property and buildings or financial instruments such as equities or bonds. Digital Asset Management and derivative terms like Brand Asset Management usually specifically relate to digital files which are usually (but not always) media related.

Audio Video Interleave (AVI)

Audio Video Interleave (AVI) is a popular multimedia format typically used for delivery of video content. AVI was invented by Microsoft in the early 1990s. Like QuickTime (a competing technology invented by Apple around the same time), AVI is known as a Container Format because it contains content that may be compressed using a variety of other codecs such as MPEG. Despite being technically inferior to a number of other formats, it has achieved a high level of market penetration and is widely supported by most video editing and playback software

Broadcast Asset Management

Broadcast Asset Management is another specialist area of Digital Asset Management and enables organizations who own (or have rights to) time-based media assets (e.g. audio or video). In a Broadcast Asset Management system, greater emphasis is placed on the ability to manage dynamic media, for example, by being able to transcode footage or append metadata to specific points in the content.

Cataloging

Cataloging means the high-level process of adding metadata to assets in a Digital Asset Management system. These are the words and terms that are part of the metadata schema used to identify an asset that enables search and filtering tools to locate it. These can be keywords, the relationship to other assets (parent–child), events, etc.

Cloud Computing

Cloud is often used as a euphemism for the Internet and Cloud Computing means services that may be offered by using multiple servers across the Internet. The main benefits of Cloud Computing are scalability, robustness, and reduced capital expenditure (for the user of cloud-based services). Scalability is made possible by adding more nodes (servers) to increase available capacity and improve performance. Robustness can be enhanced by distributing traffic across multiple nodes and providing redundancy or failover in the event one or more nodes fail. Capital expenditure can be saved when using cloud-based services since the service provider will provide the infrastructure and communications required to support their service (or build it on top of an existing cloud-based provider). Cloud Computing makes considerable use of Virtualization technology to simplify the maintenance and deployment of multiple service nodes. There are a variety of services provided via the cloud and the definition has become somewhat blurred in recent years as vendors or service providers endeavor to associate existing application services with a term which is perceived as fashionable. Some examples of Cloud Computing include the Amazon S3 storage platform, web-based application services (e.g. Google Applications), Cloud Hosting, and Content Delivery Networks (CDNs). There are a range of other cloud-based services specific to Digital Asset Management, video transcoding being a notable offering.

Codec

Codec stands for coder/decoder and refers to the encoding of analog media like audio or video into digital format and subsequent decoding upon playback. Codecs are methods of achieving this process (they are often called "algorithms"). The encoded media are sometimes referred to as essences. For practical purposes, the encoding usually means compressing the original media so it produces a file that is usable and can be stored without occupying vast amounts of storage space. Media formats for audio and video employ different codecs—generally there is an inverse relationship between the level of compression and the quality of the corresponding output.

Container Format

Container format is usually applied to multimedia digital assets and means that the file type is not a compression technology (or codec) but is used to hold media that has been encoded by other technologies. Some popular container formats include MXF, LXF, GXF, AVI, DNG, and QuickTime.

Content Delivery Network (CDN)

Content Delivery Networks or CDNs are dedicated networks with high levels of capacity specifically designed for the distribution of bandwidth heavy content. The most common use case scenario for a CDN is on-line advertising and in particular where rich media like video is utilized as part of the presentation. A CDN enables the content only to be distributed without the expense and complexity of building multiple servers. CDNs are specifically optimized for media delivery and provide services such as media streaming.

Controlled Vocabulary

Controlled vocabularies are used in indexes, subject headings, thesauri, and taxonomies. Rather than presenting a free form natural language vocabulary where any term can be supplied, controlled vocabularies offer pre-selected terms for users to choose from.

Database Server

A database server is typically used in a DAM system to hold metadata about assets. The majority of modern databases are known as *Relational Databases* (the correct term is *RDBMS [Relational Database Management System]*). In a relational database, tables of information are connected together by using identifiers (or indexes) to query them. Some examples of database servers in current use include: SQL Server, MySQL, Oracle, and Postgres.

Data Migration

Data migration is the transfer of data from one database to its replacement. After successful data migration, the original system usually ceases to be in use. Contrast with systems integration which involves the sharing of data between two live databases systems that will both remain operational.

DBMS

An abbreviation for Database Management System. See Database Server for a more detailed description.

Derivative Files

Derivative files describe assets that are created from the original. In Digital Asset Management systems, these can refer to previews that enable users to see what an asset looks like before they download it. They may include a variety of options such as thumbnail images, Flash Video, low resolution, or watermarked editions of images. As well as previews, derivative files sometimes refer to assets that will be used for production purposes but where some key aspect has been altered (e.g. the size, format, or color space). The term *derivative files* can almost be used interchangeably with surrogate files, although the former expression implies a wider range of uses.

Digital Asset Management (DAM)

Digital Asset Management (DAM) is a collective term applied to the process of storing, cataloguing, searching, and delivering computer files (or digital assets). These may take the form of video, audio, images, print marketing collateral, office documents, fonts, or 3D models. DAM systems centralize assets and establish a systematic approach to ingesting assets so they can be located more easily and used appropriately.

Digital Content Management (DCM)

Digital Content Management (DCM) is synonymous with Digital Asset Management. Although technically it specifically relates to media content as opposed to general data assets, in practical terms there is no difference between the two descriptions. The phrase Digital Content Management is often used to avoid confusion with Asset Management, which has a variety of meanings across different industries.

Digital Rights Management (DRM)

Digital Rights Management (DRM) refers to technology and practices used to protect digital intellectual property from being used in a way that breaches the terms of its license. This generally means preventing assets from being illegally copied. The term can have multiple meanings depending on whether it is being used by asset consumers or asset suppliers. In the latter case it will often imply the use of some kind of technology to prevent media from being copied from one device to another (MP3 files is particularly common), however, it can also mean controls established by media users to prevent intellectual property from being accidentally used without permission.

Digitization

Digitization is the conversion of analog or physical assets into digital equivalents. The methods for doing this are as varied as the media that a Digital Asset Management system can support. The scanning of images and conversion of film or video tends to be the most common form of digitization activity. The need to digitize assets is gradually diminishing as more media is recorded directly in digital formats.

Dublin Core Metadata Initiative (DCMI)

The Dublin Core Metadata Initiative (DCMI) is a reference to a metadata standard and the organization that first established it. Dublin Core Metadata is common in public sector Digital Asset Management systems as well as other archives and repositories of information. The aim is to provide a standardized core set of fields or criteria for the description of content (in a broad sense) as well as a framework for adding content-specific extensions. DCMI fields can be theoretically applied to almost any type of asset. DCMI data is sometimes used either in-line in the Meta tags of web pages (or as a reference to an associated XML file) as well as for other content such as photos, documents, videos, etc.

Enterprise Content Management (ECM)

Enterprise Content Management (ECM) is a wide-ranging term that is sometimes incorrectly used instead of Digital Asset Management (DAM). ECM systems tend to be large-scale repositories of many types of content held across the entirety of an organization. As well as digital media, nearly all material (including operational documents and files) may be included in the scope of an ECM implementation. The objective of providing ECM is usually to offer a single interface where employees can gain access to all of an organization's data. Many DAM systems are being integrated with ECM as an alternative method that enables organizations to leverage the benefits of both.

Encapsulated PostScript (EPS)

Encapsulated PostScript or EPS is a derivative of the PostScript standard and is a digital image format. EPS files are fully self-contained (or encapsulated) PostScript documents that come with an associated preview image so the user can view them. EPS files are more prevalent with specialist structured drawing programs such as Adobe Illustrator but are still supported by most modern desktop applications.

Essences

Essences refer to raw audio or video streams used in media files. Essences will usually be encoded with a Codec such as MPEG or MP3.

EXIF—Exchangeable Image File Format

EXIF is a metadata standard used to store information about digital images created by the Japan Electronic Industries Development Association (JEIDA, later renamed JEITA—Japan Electronics and Information Technology Industries Association) in 1998. EXIF data is usually stored inside a JPEG or TIFF file, i.e. it accompanies the image rather than being held in sidecar files (in the same fashion as IPTC and XMP). In particular, EXIF is used by manufacturers to record technical information about the digital camera used to shoot an image. XMP data offers many of the benefits of EXIF but in a more flexible and easier to manipulate fashion, however, the support for it by the digital imaging industry has ensured that EXIF remains active and in widespread use.

Flash

Flash is an application used to create ShockWave Flash (SWF) files and associated media such as Flash Video (FLV). Although the files are often referred to as "movies" they are frequently applications, interactive features, animations, or games. Flash was brought to prominence by Macromedia who acquired the original application in 1996. In Video Digital Asset Management systems, Flash movies (and FLV in particular) are often used as preview formats to allow users to check video assets before downloading them.

Flash Video (FLV)

Flash Video or FLV is a compressed video format developed specifically to allow video to be played back over the Internet via the Flash player. FLV files tend to be considerably smaller than conventional video formats which make them especially useful for previewing media prior to download in Video Digital Asset Management systems and websites that use video.

Guide File

Guide files are a type of Controlled Vocabulary where an existing file will be used to locate others. A common example is a report or brochure containing images. Users may know that a specific image was used in a document, but be unable to locate it using other search strategies. Using the document as a guide file, they can obtain a list of assets and search within this to isolate the one they require.

Hosting

Hosting refers to the process of storing and making accessible digital files or services on a remote server. In most discussions about Digital Asset Management, hosting implies that the system will be managed externally by the vendor and/or an Internet Service Provider (ISP). Hosting can be either shared between several customers of the provider or dedicated where the whole server is set aside. For hosting to be effective, there are three key components necessary: an operational server, available storage capacity to hold files, and bandwidth to send/receive requests.

ID3

ID3 is a metadata tagging standard typically used to embed metadata in MP3 audio or MP4 video. Most end users come into contact with ID3 when using software such as Apple's iTunes to catalog their collections of music files. Although popular, ID3 has a number of inherent limitations. As with the IPTC metadata standard for images, there are a fixed list of fields: title, artist, album, year, genre, and comments. The key advantage of ID3 is the widespread availability of tools that can adjust the tag data and batch process audio or video files. See the official ID3 site for more information.

Ingest

Ingest means to capture and acquire content through recording, transfer, or trans-coding from mapped locations (recording equipment, third party storage, etc.). This process often includes pertinent cataloging of metadata. Ingest may be managed from program schedule and ingest channel controller that controls all input devices, routing, and switching or may be performed via live, tape-to-file or file-to-file transfers, or other media acquisition gateways.

Interoperability

Interoperability means the ability of systems or processes to work together and is the conceptual basis of systems integration. Achieving interoperability involves two or more systems agreeing to a common protocol to exchange information. In more modern systems, this tends to be using technologies such as XML. The degree to which applications can easily integrate with each other depends on how detailed the protocol for communication is. There is a wide range of interoperability protocols used in Digital Asset Management, particularly in the area of exchanging metadata. A more common interoperability standard that has been widely adopted in the past is the Dublin Core Metadata Initiative (DCMI) schema.

JPEG

JPEG stands for Joint Photographic Experts Group, however, it more commonly refers to a compression standard that is used to reduce the disk space consumed by digital images. The compression method is referred to as "lossy" because some of the original data from the image is lost as part of the process. JPEG images are very common in Digital Asset Management solutions because they are natively supported by nearly all web browsers and their size is considerably smaller than other uncompressed formats such as TIFF (Tagged Image File Format). JPEG files are usually recognizable by the extension .jpeg or .jpg.

Keywording

Keywording is a colloquial term applied to a specific asset cataloguing activity where words, phrases, or terminology (or "keywords") are attributed to assets as metadata. Keywording is particularly relevant for photographs and images as these types of assets lack any integral descriptive information to help users identify whether they are suitable for their needs.

Lossy

"Lossy" codecs are those that compress the source media by removing (or losing) some of the information to achieve the result; MPEG is an example of a lossy codec.

Media Asset Management (MAM)

Media Asset Management (MAM) is generally considered as simply an alternative term for Digital Asset Management, although some would argue that a MAM system only supports media files rather than any type of digital file. To a greater extent, the terms are interchangeable, the expression tends to be favored when discussing Digital Asset Management for video or broadcast media contexts. In some cases, this term can refer to editorial or metadata activities associated with assets and DAM systems, for example, cataloging, keywording, or transcription of video footage or audio clips, although usually it will be called Media Asset Management Services.

Metadata

Metadata is often referred to as "data about data." In a Digital Asset Management context it refers to descriptive information applied to assets to support a task or activity. The most common example is to help users to locate assets in searches. To help find suitable media, assets will generally have short descriptions or titles added to them as a basic minimum, although it is more common to add much more descriptive detail to help users to locate what they are looking for. As well as search metadata, workflow and business process information may also be added to determine what procedures are followed when users want to download assets. There are six primary types of metadata: administrative, technical, descriptive, preservation, rights management, and structural.

- Administrative: Metadata related to the use and management of resources.
- Preservation: A form of administrative metadata documenting the preservation processes performed on resources in both conventional and digitization workflow.
- Rights Management: Metadata includes user-oriented rights information about copyright status, permissions, obligations, and restrictions pertaining to use of the asset.
- Technical: Metadata that describes the creation and technical characteristics of digital and physical objects, such as file format, resolution, size, duration, and track configuration. The automated capture of detailed technical metadata is central to obsolescence planning and preservation strategy.
- Descriptive: Metadata that describes an asset for the purposes of discovery and identification, such as titles, subjects, description, genres, and creators.
- Structural Metadata: metadata may include tape labels, card catalog cards, meeting schedule information, logs, transcripts, etc. It is possible for metadata to exist before the asset exists; for instance when a scheduled event is given a title, date, and location, then record it later, the data about the event (the metadata) was created before the creation of a recording of the actual event.

Metadata Standard

A metadata element set and/or schema that has been defined and authorized by a national or international standards body, community, or professional association. A metadata standard serves as an authority on how to define and structure metadata (i.e. SMPTE, EBUCore).

MOV (QuickTime)

MOV is the file format extension for QuickTime movies.

MPEG

MPEG stands for Moving Picture Experts Group and is a working group that develops standards for encoding digital video and audio. In the case of most Digital Asset Management systems, MPEG refers to a type of video format. There are three common variations of MPEG (named MPEG1, MPEG2, and MPEG4) along with MPEG7 and MPEG21. MPEG1 was the first standard for encoding video. MPEG2 enhanced the standard and improved support for digital storage on DVDs and other devices. MPEG3 was discontinued. MPEG4 increased the range of output devices to cover mobile and Internet delivery.

MXF

MXF stands for Material eXchange Format and is a container format for time-based media such as video and audio. MXF files allow a number of essences encoded in a given codec to be (theoretically) stored in the same file as the metadata which may be used to describe it. The implementation of MXF varies across different software systems, some will not actually use the same file to store data but rely on a single MXF header file with linked video, audio, and XML metadata stored as separate files. Despite some problems with the handling of MXF it is gaining widespread acceptance among the A/V industry as a means of archiving time-based media.

NAS (Network Attached Storage) Server

Network Attached Storage (NAS) Servers are dedicated to the storage of digital files. The purpose of having a computer whose sole purpose is file storage is to reduce the load on a web, application, or database server. Unlike an external hard disk, a NAS is usually an actual computer with an operating system installed on it. Because NAS servers are specialized toward just providing storage, extra capacity can usually be added to them easily. NAS are commonly used for Digital Asset Management projects to provide sufficient storage capacity for repositories of larger files such as video, print/artwork files, or original high resolution images. SANs (Storage Area Network) are sometimes used as an alternative to a

NAS, although this is less common with a dedicated Digital Asset Management Software.

Normalization

The term "Normalization" is generally used when designing databases to hold asset metadata. This description is highly simplified but in essence it means to index or rationalize common groups of terms down to a series of numbers so that they can be searched more quickly. Normalized data is typically found in drop-down menus or sets of checkboxes. Fully normalized data is also considerably easier to manipulate, for example, if an index or ID is used to represent a tag, changing it once will cause all assets associated with that term to be updated also. Many of the biggest issues with normalization come when migrating data from a legacy system to a new Digital Asset Management system as the older application may not be as well normalized as the new one.

Ontology

Ontology has a philosophical definition as well as an IT-oriented meaning which is more suitable in the context of Digital Asset Management. An ontology shows the relationships, properties, and functions between entities or concepts. Unlike a taxonomy, an ontology enables a wider range of relationships between attributes or terms than a simple hierarchy to be represented. This is of particular value when cataloging complex or multi-faceted asset repositories or if a DAM System is tightly integrated with Knowledge Management Systems (KMS) and Enterprise Search.

Proxy Files

This term refers to any files that are created from the original for reference purposes. They are used to represent assets—in general as a low resolution, truncated, or otherwise constrained edition. The term is now the more popular way to describe non-original assets that have been rendered specifically for use in Digital Asset Management systems. Also see Surrogate Files.

QuickTime

QuickTime is a widely adopted standard for delivery of multimedia content and was developed by Apple in the early 1990s, originally for Macintosh but Windows support was added in a later release. Although capable of dealing with other types of media such as audio, text, and 3D panoramas (such as QuickTime VR), it is generally associated with video. The QuickTime file format (see the MOV entry for more) is known as a Container Format because it holds various types of media—rather than being a native codec in its own right. The QuickTime player required to view media has a high penetration on Macintosh computers because it ships with this operating system. On Windows, it is reasonably widely deployed, however, it must be separately installed and this makes it less suitable than Flash

Video for pure web-based delivery using an online Video Digital Asset Management system.

RAW

RAW files are used by professional grade digital cameras to store images without processing them into a more common image format such as JPEG or TIFF. The characteristics of the RAW format that each camera writes tend to change depending on which component vendor a manufacturer has used for their device, this makes dealing with them using Digital Asset Management tools quite complex. The main benefit of retaining an image in RAW format is that the conversion to a more universally recognized standard tends to lose at least some information (i.e. the quality of the reproduction degrades). In this sense, RAW files can be viewed as analogous to negatives in traditional photography. Because original RAW files are by their nature precious, sidecar files are used by some applications such as Adobe Photoshop to store the changes made to them. Another format, DNG (Digital Negative) has been developed by Adobe as a Container Format for holding RAW files along with other types of data.

RDBMS

An abbreviation for Relational Database Management System. See Database Server for more information.

Really Simple Syndication (RSS)

RSS is an XML-based Metadata standard that makes it simple for websites to syndicate data from other web-based resources. RSS sources are typically referred to as "feeds" and include a short snippet about the article and a link back to it; they are common in blogs and with news sites as they allow readers to find out if there has been an update to a site without visiting it (using a "feed reader" or "aggregator"). RSS equipped Digital Asset Management systems allow new or revised assets to be published to other locations (e.g. an intranet or external website).

Real Time Streaming Protocol (RTSP)

Real Time Streaming Protocol is an Internet protocol for Media Streaming. It was developed by the Internet Engineering Task Force (IETF) to provide a basic set of commands for controlling dynamic media such as play, pause, record, etc. A variety of commercial and open source streaming products support RTSP, including Real Networks, Apple, and Microsoft via Windows Media Services.

SAN (Storage Area Network)

A SAN (Storage Area Network) is used to aggregate the storage capability available on different devices (e.g. servers) so they appear as a single disk. The key benefits of this approach are efficiency and availability. By combining storage, SANs can

prevent uneven distribution of capacity and also offer greater reliability by repli-
cating data across the network. Most SANs require special fiber optic cabling to
be effective as the performance across a conventional LAN is insufficient to be of
practical use. DAM systems tend to be established as separate facilities so the use
of a SAN is not as widespread, however, they can potentially offer some advantages
and should be considered as an option when deciding a DAM hardware and hosting strategy.

Search

There are the various ways to find assets once they are cataloged: browsing, filtered
searches, associations (parent–child). There are faceted and contextual searches
which use intuitive algorithms, and simple filtered searches using known data
points.

SWF (ShockWave Flash)

ShockWave Flash or SWF is the type of file created by the Flash application. SWF
movies are generally played back on the Flash player built into browsers, although
the format can be used on mobile devices and is sometimes embedded into other
programs also.

Sidecar Files

Sidecar files are used to hold XMP data about a RAW image. This can include
modifications to the RAW file, IPTC data, or other types of metadata. The benefit
of using sidecar files is that the metadata does not need to be contained with the
image and can be manipulated separately. The disadvantage is that this does also
mean that the metadata contained within them can become lost or divorced from
the original. Sidecar file data can also sometimes be stored in a database rather
than files to reduce the risk of loss at the expense of some flexibility.

Stemming

Stemming refers to a technique for increasing the quantity of search results by
reducing a supplied keyword search term to the base element of the word (i.e. its
stem) and then using that to try to identify other terms. For example, searching for
activation in a Digital Asset Management system that supports stemming might
yield results for *activate, actively, active, activeness*, etc.

Streaming

Streaming means the ability of media to be viewed at the same time as it is being
downloaded. The key benefit of streamed assets is that the user does not need to
wait until the entire file has been obtained before they can inspect it. There are
two basic varieties of media streaming: live and archived. Live streaming involves
capturing the output from a camera or other digital source and relaying it to

users in real-time as an event takes place. Archived streaming takes assets that have already been digitized. Streaming takes on particular significance when dealing with dynamic time-based media such as audio or video and is (to a greater extent) essential for a Video Digital Asset Management system. There are a variety of media streaming protocols in widespread use, including FLV (Flash Video), Real Time Streaming Protocol (RTSP), and 3GP (for delivery to mobile devices).

Surrogate Files

This term is now losing favor to Proxy files. Surrogate files are those derived from an original digital asset and are typically used in combination with metadata to help users locate media prior to downloading them. They usually provide a preview in the form of a thumbnail, smaller image, preview clip, or other file that can be transferred quickly. In some cases, surrogate files may be the actual file supplied, for example, if an image is to be used in a PowerPoint presentation and the user does not have a graphics program installed. Surrogate files are sometimes referred to as Derivative files. Also see watermarking for information on how surrogate files can be used to enforce copyright.

Systems Integration

Systems integration is the process of exchanging data between two or more IT systems to leverage further benefits out of the original applications. In the context of Digital Asset Management it may mean either receiving digital assets from another system (e.g. artwork from a workflow system) or providing raw data to automate an on-going business process such as providing asset ordering and pricing information to a finance system. Frequently it now refers to the process of integrating Digital Asset Management systems into enterprise-wide search tools or portals using XML. Systems integration is distinct from data migration because both systems continue to be active and co-exist semi-independently.

Tagging

Tagging is a colloquial term given to the process of adding metadata generally and keywords in particular to digital assets.

Taxonomy

Taxonomy means a classification system that is usually hierarchical in nature (i.e. it has parent-child relationships between terms). Originally a scientific term used to classify living organisms, taxonomies are now used to describe any abstract tree-like metadata structure that is composed of categories, sub-categories, and nodes. The relationship between terms is more rigid than an ontology where terms can be inter-connected using a range of polyhierarchical or non-hierarchical systems, for example, venn diagrams or matrices (note that this does not mean that an ontology is superior as a metadata structure—in many cases it is not). In Digital Asset

Management discussions, the design of a taxonomy to represent information about assets is important to enable the development of thesauri and controlled vocabularies for metadata entry and searching purposes.

Thesaurus

A thesaurus is a set of synonyms or related terms for a given word or description, unlike a taxonomy, it may be polyhierarchical and involve complex relationships such as broader or narrower terms. Thesauri describe the standard terms for concepts in a controlled vocabulary.

Transcoding

Transcoding is the process of converting one video or audio format into another. In general it refers to the conversion of one codec to another (e.g. MPEG to FLV), although the description can also apply to conversions between container formats (e.g. QuickTime to AVI).

Usage Approval

A specialized type of Digital Asset Management workflow where a user must apply before they are given the rights to download or use an asset. Typically, it will involve the proposed usage being checked manually by a human being, although, if the asset has been tagged with suitable metadata it is possible to partially automate this process by directing it to the correct person.

Watermarking

Watermarking is often used to protect assets by applying a translucent logo or image over the top of a surrogate asset such as an image, video, or document to prevent it being copied and re-used without authorization. Watermarking is very common in stock photography libraries where Digital Asset Management systems have been used to create public catalogues. It is also common in corporate Brand Asset Management systems to help enforce copyright compliance.

Workflow

Workflow refers to the modeling of the steps required to achieve a task so it can be streamlined and managed more effectively. In the asset supply chains commonly used in Digital Asset Management systems, workflow is often used at the ingestion and usage approval stages. It may also be used to integrate with artwork tracking systems to automatically publish assets after they have been originated and approved.

XBRL

XBRL stands for eXtensible Business Reporting Language and is an XML-based metadata standard for representation of business, accounting, and financial data as well as semantic relationships between these entities. As a standard it is specifically focused on financial data that would historically have to be represented by either unstructured objects (e.g. a block of text) or an arbitrary method (e.g. a spreadsheet where there is an implicit structure but not necessarily a uniform structure). The type of financial data represented is not necessarily transactional but might include information such as net profit or other aggregate information. A typical use case scenario would be for investor relations or financial publishers who wish to represent corporate accounts in a method that can be analyzed via third party applications or other automated methods (e.g. stock screening). See the official XRBL site for more details.

XML

XML is an abbreviation of eXtensible Markup Language. XML is a standard for creating markup languages which describe the structure of data so that it can be exchanged between two different systems. It is heavily used in systems integration. Most second generation Digital Asset Management (DAM) systems include features that allow metadata and assets to be supplied to third party systems in XML format. More advanced Digital Asset Management systems also allow third party applications to integrate with them using XML web services.

XMP

XMP is an abbreviation of eXtensible Metadata Platform and is a form of XML and is a metadata standard for describing assets such as images and documents. XMP is widely regarded as the successor to IPTC as it allows the range of metadata fields used to describe assets to be extended as required.

Index

Page numbers in *italics* refer to figures and tables.